Deep Reinforcement Learning for Wireless Communications and Networking

Deep Reinforcement Learning for Wireless Communications and Networking

Theory, Applications, and Implementation

Dinh Thai Hoang
University of Technology Sydney, Australia

Nguyen Van Huynh
Edinburgh Napier University, United Kingdom

Diep N. Nguyen
University of Technology Sydney, Australia

Ekram Hossain
University of Manitoba, Canada

Dusit Niyato
Nanyang Technological University, Singapore

IEEE PRESS

WILEY

Published by John Wiley & Sons, Inc., Hoboken, New Jersey.
Published simultaneously in Canada.

For general information on our other products and services or for technical support, please contact our Customer Care Department within the United States at (800) 762-2974, outside the United States at (317) 572-3993 or fax (317) 572-4002.

Wiley also publishes its books in a variety of electronic formats. Some content that appears in print may not be available in electronic formats. For more information about Wiley products, visit our web site at www.wiley.com.

Library of Congress Cataloging-in-Publication Data applied for:

Hardback ISBN: 9781119873679

Cover Design: Wiley
Cover Image: © Liu zishan/Shutterstock

Set in 9.5/12.5pt STIXTwoText by Straive, Chennai, India

To my family – Dinh Thai Hoang
To my family – Nguyen Van Huynh
To Veronica Hai Binh, Paul Son Nam, and Thuy – Diep N. Nguyen
To my parents – Ekram Hossain
To my family – Dusit Niyato

Contents

Notes on Contributors

Dinh Thai Hoang
School of Electrical and Data
Engineering
University of Technology Sydney
Australia

Nguyen Van Huynh
School of Computing, Engineering and
the Built Environment
Edinburgh Napier University
UK

Diep N. Nguyen
School of Electrical and Data
Engineering
University of Technology Sydney
Australia

Ekram Hossain
Department of Electrical and
Computer Engineering
University of Manitoba
Canada

Dusit Niyato
School of Computer Science and
Engineering
Nanyang Technological University
Singapore

Foreword

Prof. Merouane Debbah, Integrating deep reinforcement learning (DRL) techniques in wireless communications and networking has paved the way for achieving efficient and optimized wireless systems. This ground-breaking book provides excellent material for researchers who want to study applications of deep reinforcement learning in wireless networks, with many practical examples and implementation details for the readers to practice. It also covers various topics at different network layers, such as channel access, network slicing, and content caching. This book is essential for anyone looking to stay ahead of the curve in this exciting field.

Prof. Vincent Poor, Many aspects of wireless communications and networking are being transformed through the application of deep reinforcement learning (DRL) techniques. This book represents an important contribution to this field, providing a comprehensive treatment of the theory, applications, and implementation of DRL in wireless communications and networking. An important aspect of this book is its focus on practical implementation issues, such as system design, algorithm implementation, and real-world deployment challenges. By bridging the gap between theory and practice, the authors provide readers with the tools to build and deploy DRL-based wireless communication and networking systems. This book is a useful resource for those interested in learning about the potential of DRL to improve wireless communications and networking systems. Its breadth and depth of coverage, practical focus, and expert insights make it a singular contribution to the field.

Preface

Reinforcement learning is one of the most important research directions of machine learning (ML), which has had significant impacts on the development of artificial intelligence (AI) over the last 20 years. Reinforcement learning is a learning process in which an agent can periodically make decisions, observe the results, and then automatically adjust its strategy to achieve an optimal policy. However, this learning process, even with proven convergence, often takes a significant amount of time to reach the best policy as it has to explore and gain knowledge of an entire system, making it unsuitable and inapplicable to large-scale systems and networks. Consequently, applications of reinforcement learning are very limited in practice. Recently, deep learning has been introduced as a new breakthrough ML technique. It can overcome the limitations of reinforcement learning and thus open a new era for the development of reinforcement learning, namely ***deep reinforcement learning*** (DRL). DRL embraces the advantage of deep neural networks (DNNs) to train the learning process, thereby improving the learning rate and the performance of reinforcement learning algorithms. As a result, DRL has been adopted in numerous applications of reinforcement learning in practice such as robotics, computer vision, speech recognition, and natural language processing.

In the areas of communications and networking, DRL has been recently used as an effective tool to address various problems and challenges. In particular, modern networks such as the Internet-of-Things (IoT), heterogeneous networks (HetNets), and unmanned aerial vehicle (UAV) networks become more decentralized, ad-hoc, and autonomous in nature. Network entities such as IoT devices, mobile users, and UAVs need to make local and independent decisions, e.g. spectrum access, data rate adaption, transmit power control, and base station association, to achieve the goals of different networks including, e.g. throughput maximization and energy consumption minimization. In uncertain and stochastic environments, most of the decision-making problems can be modeled as a so-called *Markov decision process* (MDP). Dynamic programming and other algorithms such as value iteration, as well as reinforcement learning techniques, can be adopted to solve the MDP. However, modern networks are large-scale and complicated, and thus the computational complexity of the techniques rapidly becomes unmanageable, i.e. curse of dimensionality. As a result, DRL has been developing as an alternative solution to overcome the challenge. In general, the DRL approaches provide the following advantages:

- DRL can effectively obtain the solution of sophisticated network optimizations, especially in cases with incomplete information. Thus, it enables network entities, e.g. base stations, in modern networks to solve non-convex and complex problems, e.g. joint user association, computation, and transmission schedule, to achieve optimal solutions without complete and accurate network information.
- DRL allows network entities to learn and build knowledge about the communication and networking environment. Thus, by using DRL, the network entities, e.g. a mobile user, can learn optimal policies, e.g. base station selection, channel selection, handover decision, caching, and offloading decisions, without knowing a priori channel model and mobility pattern.
- DRL provides autonomous decision-making. With the DRL approach, network entities can make observations and obtain the best policy locally with minimum or without information exchange among each other. This not only reduces communication overheads but also improves the security and robustness of the networks.
- DRL significantly improves the learning speed, especially in problems with large state and action spaces. Thus, in large-scale networks, e.g. IoT systems with thousands of devices, DRL allows the network controller or IoT gateways to control dynamically user association, spectrum access, and transmit power for a massive number of IoT devices and mobile users.
- Several other problems in communications and networking such as cyber-physical attacks, interference management, and data offloading can be modeled as games, e.g. the non-cooperative game. DRL has been recently extended and used as an efficient tool to solve competitor, e.g. finding the Nash equilibrium, without complete information.

Clearly, DRL will be the key enabler for the next generation of wireless networks. Therefore, DRL is of increasing interest to researchers, communication engineers, computer scientists, and application developers. In this regard, we introduce a new book, titled "***Deep Reinforcement Learning for Wireless Communications and Networking: Theory, Applications, and Implementation***", which will provide a fundamental background of DRL and then study recent advances in DRL to address practical challenges in wireless communications and networking. In particular, this book first gives a tutorial on DRL, from basic concepts to advanced modeling techniques to motivate and provide fundamental knowledge for the readers. We then provide case studies together with implementation details to help the readers better understand how to practice and apply DRL to their problems. After that, we review DRL approaches that address emerging issues in communications and networking. The issues include dynamic network access, data rate control, wireless caching, data offloading, network security, and connectivity preservation, which are all important to next-generation networks such as 5G and beyond. Finally, we highlight important challenges, open issues, and future research directions for applying DRL to wireless networks.

Acknowledgments

The authors would like to acknowledge grant-awarding agencies that supported parts of this book. This research was supported in part by the Australian Research Council under the DECRA project DE210100651 and the Natural Sciences and Engineering Research Council of Canada (NSERC).

The authors would like to thank Mr. Cong Thanh Nguyen, Mr. Hieu Chi Nguyen, Mr. Nam Hoai Chu, and Mr. Khoa Viet Tran for their technical assistance and discussions during the writing of this book.

Acronyms

No	Acronyms	Terms
1	A3C	asynchronous advantage actor-critic
2	ACK	acknowledgment message
3	AI	artificial intelligence
4	ANN	artificial neural network
5	AP	access point
6	BER	bit error rate
7	BS	base station
8	CNN	convolutional neural network
9	CSI	channel state information
10	D2D	device-to-device
11	DDPG	deep deterministic policy gradient
12	DDQN	double deep Q-network
13	DL	deep learning
14	DNN	deep neural network
15	DPG	deterministic policy gradient
16	DQN	deep Q-learning
17	DRL	deep reinforcement learning
18	eMBB	enhanced mobile broadband
19	FL	federated learning
20	FSMC	finite-state Markov chain
21	GAN	generative adversarial network
22	GPU	graphics processing unit
23	IoT	Internet-of-Things

No	Acronyms	Terms
24	ITS	intelligent transportation system
25	LTE	Long-term evolution
26	M2M	machine-to-machine
27	MAC	medium access control
28	MARL	multi-agent RL
29	MDP	Markov decision process
30	MEC	mobile edge computing
31	MIMO	multiple-input multiple-output
32	MISO	Multi-input single-output
33	ML	machine learning
34	mMTC	massive machine type communications
35	mmWave	millimeter wave
36	MU	mobile user
37	NFV	network function virtualization
38	OFDMA	orthogonal frequency division multiple access
39	POMDP	partially observable Markov decision process
40	PPO	proximal policy optimization
41	PSR	predictive state representation
42	QoE	Quality of Experience
43	QoS	Quality of Service
44	RAN	radio access network
45	RB	resource block
46	RF	radio frequency
47	RIS	reconfigurable intelligent surface
48	RL	reinforcement learning
49	RNN	recurrent neural network
50	SARSA	state-action-reward-state-action
51	SDN	software-defined networking
52	SGD	stochastic gradient descent
53	SINR	signal-to-interference-plus-noise ratio
54	SMDP	semi-Markov decision process
55	TD	temporal difference
56	TDMA	time-division multiple access
57	TRPO	trust region policy optimization

No	Acronyms	Terms
58	UAV	unmanned aerial vehicle
59	UE	user equipment
60	UL	uplink
61	URLLC	ultra-reliable and low-latency communications
62	VANET	vehicular ad hoc NETworks
63	VNF	virtual network function
64	WLAN	wireless local area network
65	WSN	wireless sensor network

Introduction

Deep reinforcement learning (DRL) empowered by deep neural networks (DNNs) has been developing as a promising solution to address high-dimensional and continuous control problems effectively. The integration of DRL into future wireless networks will revolutionize conventional model-based network optimization with model-free approaches and meet various application demands. By interacting with the environment, DRL provides an autonomous decision-making mechanism for the network entities to solve non-convex, complex, model-free problems, e.g. spectrum access, handover, scheduling, caching, data offloading, and resource allocation. This not only reduces communication overhead but also improves network security and reliability. Though DRL has shown great potential to address emerging issues in complex wireless networks, there are still domain-specific challenges that require further investigation. The challenges may include the design of proper DNN architectures to capture the characteristics of 5G network optimization problems, the state explosion in dense networks, multi-agent learning in dynamic networks, limited training data and exploration space in practical networks, the inaccessibility and high cost of network information, as well as the balance between information quality and learning performance.

This book provides a comprehensive overview of DRL and its applications to wireless communication and networking. It covers a wide range of topics from basic to advanced concepts, focusing on important aspects related to algorithms, models, performance optimizations, machine learning, and automation for future wireless networks. As a result, this book will provide essential tools and knowledge for researchers, engineers, developers, and graduate students to understand and be able to apply DRL to their work. We believe that this book will not only be of great interest to those in the fields of wireless communication and networking but also to those interested in DRL and AI more broadly.

Part I

Fundamentals of Deep Reinforcement Learning

1

Deep Reinforcement Learning and Its Applications

1.1 Wireless Networks and Emerging Challenges

Over the past few years, communication technologies have been rapidly developing to support various aspects of our daily lives, from smart cities and healthcare to logistics and transportation. This will be the backbone for the future's data-centric society. Nevertheless, these new applications generate a tremendous amount of workload and require high-reliability and ultrahigh-capacity wireless communications. In the latest report [1], Cisco projected the number of connected devices that will be around 29.3 billion by 2023, with more than 45% equipped with mobile connections. The fastest-growing mobile connection type is likely machine-to-machine (M2M), as Internet-of-Things (IoT) services play a significant role in consumer and business environments. This poses several challenges in future wireless communication systems:

- Emerging services (e.g. augmented reality [AR] and virtual reality [VR]) require high-reliability and ultrahigh capacity wireless communications. However, existing communication systems, designed and optimized based on conventional communication theories, significantly prevent further performance improvements for these services.
- Wireless networks are becoming increasingly ad hoc and decentralized, in which mobile devices and sensors are required to make independent actions such as channel selections and base station associations to meet the system's requirements, e.g. energy efficiency and throughput maximization. Nonetheless, the dynamics and uncertainty of the systems prevent them from obtaining optimal decisions.
- Another crucial component of future network systems is network traffic control. Network control can dramatically improve resource usage and the efficiency of information transmission through monitoring, checking, and controlling data flows. Unfortunately, the proliferation of smart IoT devices and ultradense

Deep Reinforcement Learning for Wireless Communications and Networking:
Theory, Applications, and Implementation, First Edition.
Dinh Thai Hoang, Nguyen Van Huynh, Diep N. Nguyen, Ekram Hossain, and Dusit Niyato.
© 2023 The Institute of Electrical and Electronics Engineers, Inc. Published 2023 by John Wiley & Sons, Inc.

radio networks has greatly expanded the network size with extremely dynamic topologies. In addition, the explosive growing data traffic imposes considerable pressure on Internet management. As a result, existing network control approaches may not effectively handle these complex and dynamic networks.

- Mobile edge computing (MEC) has been recently proposed to provide computing and caching capabilities at the edge of cellular networks. In this way, popular contents can be cached at the network edge, such as base station, end-user devices, and gateways to avoid duplicate transmissions of the same content, resulting in better energy and spectrum usage [2, 3]. One major challenge in future communication systems is the straggling problems at both edge nodes and wireless links, which can significantly increase the computation delay of the system. Additionally, the huge data demands of mobile users and the limited storage and processing capacities are critical issues that need to be addressed.

Conventional approaches to addressing the new challenges and demands of modern communication systems have several limitations. First, the rapid growth in the number of devices, the expansion of network scale, and the diversity of services in the new era of communications are expected to significantly increase the amount of data generated by applications, users, and networks [1]. However, traditional solutions may be unable to process and utilize this data effectively to improve system performance. Second, existing algorithms are not well-suited to handle the dynamic and uncertain nature of network environments, resulting in poor performance [4]. Finally, traditional optimization solutions often require complete information about the system to be effective, but this information may not be readily available in practice, limiting the applicability of these approaches. Deep reinforcement learning (DRL) has the potential to overcome these limitations and provide promising solutions to these challenges.

DRL leverages the benefits of deep neural networks (DNNs), which have proven effective in tackling complex, large-scale engines, speech recognition, medical diagnosis, and computer vision. This makes DRL well suited for managing the increasing complexity and scale of future communication networks. Additionally, DRL's online deployment allows it to effectively handle the dynamics and unpredictable nature of wireless communication environments.

1.2 Machine Learning Techniques and Development of DRL

1.2.1 Machine Learning

Machine learning (ML) is a problem-solving paradigm where a machine learns a particular task (e.g. image classification, document text classification, speech recognition, medical diagnosis, robot control, and resource allocation in

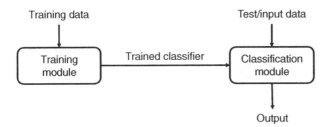

Figure 1.1 A data-driven ML architecture.

communication networks) and performance metric (e.g. classification accuracy and performance loss) using experiences or data [5]. The task generally involves a function that maps well-defined inputs to well-defined outputs. The essence of data-driven ML is that there is a pattern in the task inputs and the outcome which cannot be pinned down mathematically. Thus, the solution to the task, which may involve making a decision or predicting an output, cannot be programmed explicitly. If the set of rules connecting the task inputs and output(s) were known, a program could be written based on those rules (e.g. if-then-else codes) to solve the problem. Instead, an ML algorithm learns from the input data set, which specifies the correct output for a given input; that is, an ML method will result in a program that uses the data samples to solve the problem. A data-driven ML architecture for the classification problem is shown in Figure 1.1. The training module is responsible for optimizing the classifier from the training data samples and providing the classification module with a trained classifier. The classification module determines the output based on the input data. The training and classification modules can work independently. The training procedure generally takes a long time. However, the training module is activated only periodically. Also, the training procedure can be performed in the background, while the classification module operates as usual.

There are three categories of ML techniques, including supervised, unsupervised, and reinforcement learning.

- *Supervised learning*: Given a data set $\mathbf{D} = \{(\mathbf{x}_1, y_1), (\mathbf{x}_2, y_2), \ldots, (\mathbf{x_n}, y_n)\} \subseteq \mathbb{R}^n \times C$, a supervised learning algorithm predicts y that generalizes the input–output mapping in $\mathbf{D}$ to inputs $\mathbf{x}$ outside $\mathbf{D}$. Here, $\mathbb{R}^n$ is the n-dimensional feature space C, $\mathbf{x_i}$ is the input vector of the ith sample, y_i is the label of the ith sample, and C is the label space. For binary classification problems (e.g. spam filtering), $C = \{0,1\}$ or $C = \{-1,1\}$. For multiclass classification (e.g. face classification), $C = \{1,2,\ldots,K\}(K \geq 2)$. On the other hand, for regression problems (e.g. predicting temperature), $C = \mathbb{R}$. The data points $(\mathbf{x_i}, y_i)$ are drawn from a (unknown) distribution $\mathcal{P}(X, Y)$. The learning process involves learning a function h such that for a new pair $(\mathbf{x}, y) \backsim \mathcal{P}$, we have $h(\mathbf{x}) = y$ with high probability (or $h(\mathbf{x}) \approx y$). A loss function (or risk function), such as the mean squared error function, evaluates the error between the predicted probabilities/values returned by the function $h(\mathbf{x_i})$ and the labels y_i on the training data.

For supervised learning, the data set $\mathbf{D}$ is usually split into three subsets: $\mathbf{D}_{TR}$ as the training data, $\mathbf{D}_{VA}$ as the validation data, and $\mathbf{D}_{TE}$ as the test data. The function $h(\cdot)$ is validated on $\mathbf{D}_{VA}$: if the loss is too significant, $h(\cdot)$ will be revised based on $\mathbf{D}_{TR}$ and validated again on $\mathbf{D}_{VA}$. This process will keep going back and forth until it gives a low loss on $\mathbf{D}_{VA}$. The standard supervised learning techniques include the following: Bayesian classification, logistic regression, K-nearest neighbor (KNN), neural network (NN), support vector machine (SVM), decision tree (DT) classification, and recommender system. Note that supervised learning techniques require the availability of labeled data sets.

- *Unsupervised learning* techniques are used to create an internal representation of the input, e.g. to form clusters, extract features, reduce dimensionality, estimate density. Unlike supervised learning, these techniques can deal with unlabeled data sets.
- *Reinforcement learning (RL)* techniques do not require a prior dataset. With RL, an agent learns from interactions with an external environment. The idea of learning by interacting with a domain is an imitation of humans' natural learning process. For example, at the point when a newborn child plays, e.g. waves his arms or kicks a ball, his/her brain has a direct sensorimotor connection with its surroundings. Repeating this process produces essential information about the impact of actions, causes and effects, and what to do to reach the goals.

Deep learning (DL), a subset of ML, has gained popularity thanks to its DNN architectures to overcome the limitations of ML. DL models are able to extract the key features of data without relying on the data's structure. The "deep" in deep learning refers to the number of layers in the DNN architecture, with more layers leading to a deeper network. DL has been successfully applied in various fields, including face and voice recognition, text translation, and intelligent driver assistance systems. It has several advantages over traditional algorithms as follows [6]:

- *No need for system modeling*: The system must be well modeled in traditional optimization approaches to obtain the optimal solution. Nevertheless, all information about the system must be available to formulate the optimization problem. In practice, this may not be feasible, especially in future wireless networks where users' behaviors and network states are diverse and may randomly occur. Even if the optimization problem is well defined, solving it is usually challenging due to nonconvexity and high-dimensional problems. DL can efficiently address all these issues by allowing us to be data-driven. In particular, it obtains the optimal solution by training the DNN with sufficient data.
- *Supports parallel and distributed algorithms*: In many complex systems, DL may require a large volume of labeled data to train its DNN to achieve good training

performance. DL can be implemented in parallel and distributed to accelerate the training process. Specifically, instead of training with single computing hardware (e.g. graphics processing unit [GPU]), we can simultaneously leverage the computing power of multiple computers/systems for the training process. There are two types of parallelism in DL: (i) model parallelism and (ii) data parallelism. For the former, different layers in the deep learning model can be trained in parallel on other computing devices. The latter uses the same model for every execution unit but trains the model with different training samples.

- *Reusable*: The trained model can be reused in other systems/problems effectively with DL. Using well-trained models built by experts can significantly reduce the training time and related costs. For example, AlexNet can be reused in new recognition tasks with minimal configurations [6]. Moreover, the trained model can be transferred to a different but related system to improve its training using the transfer learning technique. The transfer learning technique can obtain a good training accuracy for the target system with a few training samples as it can leverage the gained knowledge in the source system. This is very helpful as collecting training samples is costly and requires human intervention.

There are several types of DNNs, such as artificial neural networks (ANNs) (i.e. feed-forward neural networks), convolutional neural networks (CNNs), and recurrent neural networks (RNNs). However, they consist of the same components: (i) neurons, (ii) weights, (iii) biases, and (iv) functions. Typically, layers are interconnected via nodes (i.e. neurons) in a DNN. Each neuron has an activation function to compute the output given the weighted inputs, i.e. synapses and bias [7]. During the training, neural network parameters are updated by calculating the gradient of the loss function.

1.2.2 Artificial Neural Network

An ANN is a typical neural network known as a feed-forward neural network. In particular, ANN consists of nonlinear processing layers, including an input layer, several hidden layers, and an output layer, as illustrated in Figure 1.2. A hidden layer uses the outputs of its previous layer as the input. In other words, ANN passes information in one direction from the input layer to the output layer. In general, ANN can learn any nonlinear function; thus, it is often referred to as a universal function approximator. The essential component of this universal approximation is activation functions. Specifically, these activation functions introduce nonlinear properties to the network and thus help it to learn complex relationships between input data and their outputs. In practice, there are three main activation functions widely adopted in DL applications: (i) sigmoid, (ii) tanh, and (iii) relu [6, 8]. Due to its effectiveness and simplicity, ANN is the most popular neural network used in DL applications.

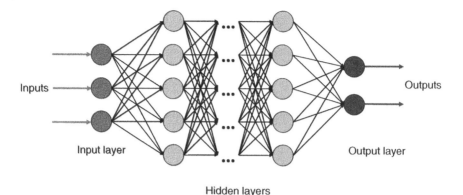

Figure 1.2 Artificial neural network architecture.

1.2.3 Convolutional Neural Network

Another type of deep neural network is CNN, designed mainly to handle image data. To do that, CNN introduces new layers, including convolution, Rectified Linear unit (Relu), and pooling layers, as shown in Figure 1.3.

- *Convolution layer* deploys a set of convolutional filters, each of which handles certain features from the images.
- *Relu layer* can map negative values to zero and maintain positive values during training, and thus it enables faster and more effective training.
- *Pooling layer* is designed to reduce the number of parameters that the network needs to learn by performing down-sampling operations.

It is worth noting that a CNN can contain tens or hundreds of layers depending on the given problem. The filters can learn simple features such as brightness and edges and then move to complex properties that uniquely belong to the object. In general, CNN performs much better than ANN in handling image data. The main reason is that CNN does not need to convert images to one-dimensional vectors before training the model, which increases the number of trainable parameters and cannot capture the spatial features of images. In contrast, CNN uses convolutional layers to learn the features of images directly. As a result, it can effectively

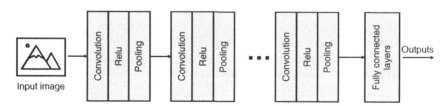

Figure 1.3 Convolutional neural network architecture.

learn all the features of input images. In the area of wireless communications, CNN is a promising technique to handle network data in the form of images, e.g. spectrum analysis [9–11], modulation classification [12, 13], and wireless channel feature extraction [14].

1.2.4 Recurrent Neural Network

An RNN is a DL network structure that leverages previous information to improve the learning process for the current and future input data. To do that, RNN is equipped with loops and hidden states. As illustrated in Figure 1.4, by using the loops, RNN can store previous information in the hidden state and operate in sequence. In particular, the output of the RNN cell at time $t - 1$ will be stored in the hidden state h_t and will be used to improve the training process of the input at time t. This unique property makes RNN suitable for dealing with sequential data such as natural language processing and video analysis.

In practice, RNN may not perform well with learning long-term dependencies as it can encounter the "vanishing" or "exploding" gradient problem caused by the backpropagation operation. Long short-term memory (LSTM) is proposed to deal with this issue. As illustrated in Figure 1.5, LSTM uses additional gates to decide

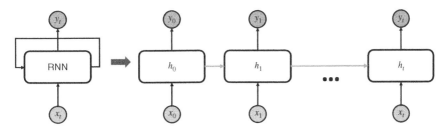

Figure 1.4 Recurrent neural network architecture.

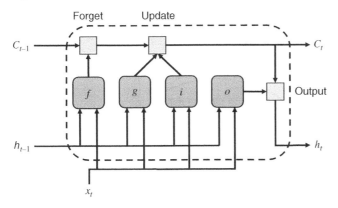

Figure 1.5 LSTM network architecture.

the proportion of previous information in the hidden state for the output and the next hidden state. Recently, RNN, especially LSTM, has emerged as a prominent structure for signal classification tasks in wireless communications [15–18]. The reason is that signals are naturally sequential data since they are usually collected over time.

1.2.5 Development of Deep Reinforcement Learning

In the last 20 years, RL [19] has become one of the most important lines of research in ML and has had an impact on the development of artificial intelligence (AI). An RL process entails an agent making regular decisions, observing the results, and then automatically adjusting their strategy to achieve the optimal policy to maximize the system performance, as illustrated in Figure 1.6. In particular, given a state, the agent makes an action and observes the immediate reward and next state of the environment. Then, this experience (i.e. current state, current action, immediate reward, and next state) will be used to update the Q-table using the Bellman equation to obtain the optimal policy. By interacting with the environment, RL can effectively deal with the dynamics and uncertainty of the environment. When the environment changes, the algorithm can adapt to the new properties and obtain a new optimal policy. While traditional RL techniques are limited to low-dimensional problems, DRL techniques are very effective to handle large-dimensional issues. In particular, DRL is a combination of RL and a DNN, where the DNN generates an action as the output given the

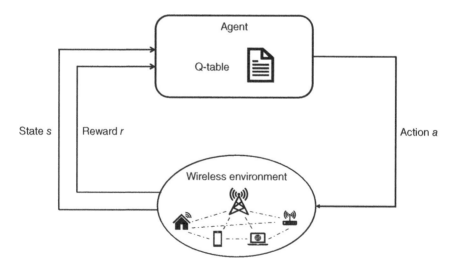

Figure 1.6 An illustration of a reinforcement learning process.

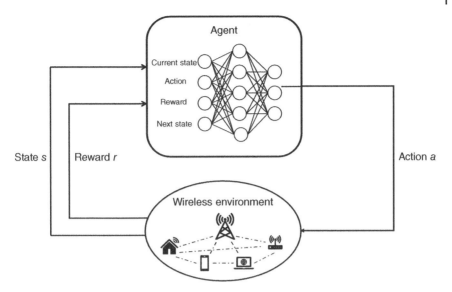

Figure 1.7 An illustration of a DRL process.

current state of the system as the input, as illustrated in Figure 1.7. This unique feature of DRL is beneficial in future wireless communication systems in which users and network devices are diverse and dynamic.

1.3 Potentials and Applications of DRL

The applications of DRL have the potential to attract significant attention from the research community and industry. The reasons for this can be explained based on the following observations.

1.3.1 Benefits of DRL in Human Lives

According to the McKinsey report, AI techniques, including DL and RL, can create between US$ 3.5 trillion and US$ 5.8 trillion annually across nine business functions in 19 industries. Experts believe DRL is at the cutting edge and has finally been applied practically to real-world applications. They also believe that the development of DRL will significantly impact AI advancement and can eventually bring us closer to artificial general intelligence (AGI).

1.3.2 Features and Advantages of DRL Techniques

DRL inherits outstanding advantages of both RL and DL techniques, thereby offering distinctive features which are expected to address and overcome diverse technical challenges existing for many years [7, 20].

- *Ability to solve sophisticated optimizations*: DRL can solve complex optimization problems by simply learning from its interactions with the system. As a result, it allows network controllers (e.g. gateways, base stations, and service providers) to solve nonconvex optimization problems to maximize the system performance in certain aspects without requiring complete and accurate network properties in advance.
- *Intelligent learning*: By interacting with the communication and networking environment, DRL can intelligently adapt its optimal policy corresponding to the system's conditions and properties. Thus, by using DRL, network controllers can efficiently perform optimal actions (e.g. beam associations, handover decisions, and caching and offloading decisions) in a real-time manner.
- *Enable fully autonomous systems*: With DRL, network entities can automatically and intelligently make optimal control decisions with minimal or without information exchange by interacting with the environment (e.g. wireless connections, users, and physical infrastructures). For that, DRL can greatly reduce communication overhead as well as enhance the system's security and robustness.
- *Overcome limitations of conventional ML techniques*: Compared with conventional RL approaches, DRL has a much faster convergence speed, especially in large-scale problems with large action and state spaces (e.g. IoT networks that contain a massive number of devices with different configurations). As a result, with DRL, network controllers can quickly adapt with the new conditions of the network and adjust their policies (e.g. user association, transmit power, and spectrum access) to maximize the long-term performance of the system.
- *Novel solutions to several conventional optimization problems*: Various problems in communications and networking, such as data offloading, interference control, and cyber-security, can be modeled as optimization problems. Unfortunately, conventional approaches cannot efficiently solve these problems due to their nonconvexity and the lack of global information. DRL, on the other hand, can effectively address these issues by simply interacting with the system to learn and obtain the optimal solutions, thanks to the power of DNNs.

1.3.3 Academic Research Activities

A substantial amount of research activities related to DRL in communications and wireless networks have been initiated. These days, the major flagship IEEE conferences (e.g. IEEE International Conference on Communications [ICC], IEEE

Global Communications Conference [GLOBECOM], IEEE Wireless Communications and Networking Conference [WCNC], and IEEE Vehicular Technology Conference [VTC]) have special sessions on the DRL techniques. Some recent IEEE magazines and journals have had their special issues on this topic, e.g. IEEE Transactions on Cognitive Communications and Networking's special issue on "Deep Reinforcement Learning for Future Wireless Communication Networks" in 2019 and IEEE IoT Journals' special issue on "Deep Reinforcement Learning for Emerging IoT Systems" in 2019. Recently, there are quite many tutorials on this topic presented in IEEE flagship conferences, such as IEEE GLOBECOM, IEEE ICC, and IEEE WF-IoT. Clearly, the research on the DRL in wireless networks is emerging and has already received significant attention from researchers worldwide.

1.3.4 Applications of DRL Techniques

As mentioned, DRL can significantly improve the learning process of reinforcement learning algorithms by using DNNs. As a result, there are many practical applications of DRL in various areas such as robotics, manufacturing, and healthcare as follows [7, 20]:

- *Video games*: In complex interactive video games, DRL is utilized to enable the agent to adapt its behavior based on its learning from the game in order to maximize the score. A wide range of PC games, such as StartCraft, Chess, Go, and Atari, utilize DRL to enable enemies to adapt their moves and tactics based on the human player's performance, as illustrated in Figure 1.8.

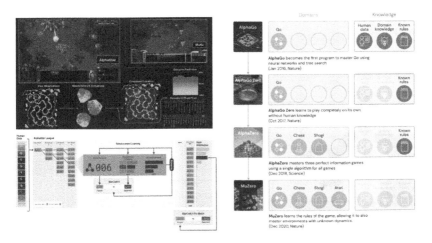

Figure 1.8 Google DeepMind's DRL applications in playing games. Adapted from [21, 22].

- *Chemistry*: DRL has been applied to optimize chemical reactions by using an agent to predict the actions that will lead to the most desirable chemical reaction at each stage of the experiment. In many cases, DRL has been found to outperform traditional algorithms used for this purpose.
- *Manufacturing*: DRL is used by many manufacturing companies, such as Fanuc, to assist their robots in picking up objects from one box and placing them in a container with high speed and accuracy [23]. These DRL-assisted machines are able to learn and memorize the objects they handle, allowing them to perform tasks efficiently and effectively [24–26]. In warehouses and e-commerce facilities, these intelligent robots are utilized to sort and deliver products to customers. For example, Tesla's factories used DRL to reduce the risk of human error by performing a significant portion of the work on vehicles.
- *Robotics*: Google developed the Soft actor-Critic algorithm, which allows robots to learn real-world tasks using DRL efficiently and safely, without the need for numerous attempts, as depicted in Figure 1.9. The algorithm has been successful in quickly training an insect-like robot to walk and a robot hand to perform simple tasks. It helps protect the robot from taking actions that could potentially cause harm.
- *Healthcare*: DRL can be used on historical medical data to determine the most effective treatments and to predict the best treatment options for current patients. For instance, DRL has been applied to predict drug doses for sepsis patients, identify optimal cycles for chemotherapy, and select dynamic treatment regimens combining hundreds of medications based on medical registry data [28–30].

In addition, many other real-world applications can be summarized in Figure 1.10.

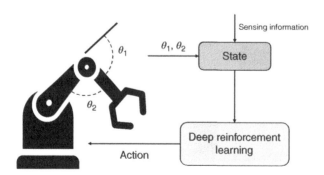

Figure 1.9 Applications of DRL in robotics. Adapted from Pilarski et al. [27].

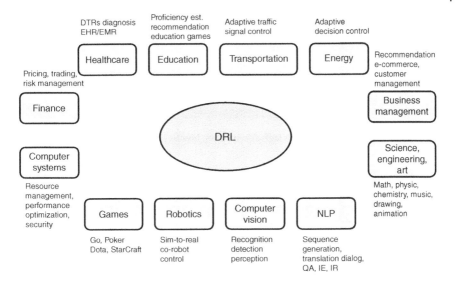

Figure 1.10   Real-word applications of DRL. EHR; electronic health record, EMR; electronic medical record, NLP; natural language processing, QA; question answering, IE; information extraction, IR; information retrieval.

1.3.5   Applications of DRL Techniques in Wireless Networks

With many applications of DRL in practice, especially for wireless and mobile networks, many patents for DRL techniques have been issued in recent years. KR102240442B1 patent [31] claims a novel idea of using DRL for proactive caching in millimeter-wave vehicular networks. CN109474980B invention [32] provides a wireless network resource allocation method based on DRL, which can improve the energy efficiency in a time-varying channel environment to the maximum extent with lower complexity. CN110488861B invention [33] discloses an unmanned aerial vehicle (UAV) track optimization method and device based on DRL and an UAV. More importantly, many applications of DRL have been deployed in our real lives, such as applications of DRL for self-driving cars and industrial automation. AWS DeepRacer is an example of an autonomous racing car that utilizes DRL to navigate a physical track using cameras and an RL model to control its throttle and direction. Similarly, Wayve.ai has recently used DRL to train a car to drive within a single day by using an algorithm to complete a lane following task using a deep network with four convolutional layers and three fully connected layers. The example in Figure 1.11 demonstrates the lane following task from the perspective of the driver. Obviously, the development of DRL will be the next big thing for future wireless networks.

Figure 1.11 DRL applications in self-driving cars. Mwiti [34].

1.4 Structure of this Book and Target Readership

1.4.1 Motivations and Structure of this Book

DRL has been shown to be a promising approach for addressing high-dimensional and continuous control problems using deep neural networks as powerful function approximators. The integration of DRL into future wireless networks will transform the traditional model-based network optimization approach to a model-free approach and enable the network to meet a variety of application requirements. By interacting with the environment, DRL provides a mechanism for network entities to make autonomous decisions and solve complex, model-free problems such as spectrum access, handover, scheduling, caching, data offloading, and resource allocation [35–37]. This can not only reduce communication overhead but also improve network security and reliability. While DRL has demonstrated significant potential for addressing emerging issues in complex wireless networks, there are still domain-specific challenges that require further investigation, such as designing appropriate DNN architectures for future network optimization problems, addressing state explosion in dense networks, handling multiagent learning in dynamic networks, dealing with limited training data and exploration space in practical networks, and balancing the trade-off between information quality and learning performance.

There are five primary objectives of this book.

- Introduce an emerging research topic together with promising applications of DRL in wireless networks.
- Provide fundamental knowledge, including comprehending theory, building system models, formulating optimization problems, and designing appropriate algorithms to address practical problems in wireless communications and networks using DRL.
- Provide a short tutorial to help the readers learn and practice programming DRL under a specific scenario.

- Provide a comprehensive review of the state-of-the-art research and development covering different aspects of DRL in wireless communications and networks.
- Introduce emerging applications of DRL in wireless communications and networks and highlight their challenges and open issues.

To achieve the objectives above, the book includes three main parts as follows:

Part I: Fundamentals of Deep Reinforcement Learning
This part presents an overview of the development of DRL and provides fundamental knowledge about theories, formulation, design, learning models, algorithms, and implementation of DRL, together with a particular case study to practice.

Chapter 1: The first chapter provides an overview of DRL, its development, and potential applications. In particular, the chapter starts with the development of wireless networks and the emerging challenges that researchers and practitioners face. We then present the remarkable development of ML and its significant impacts on all aspects of our lives. After that, we introduce recent breakthroughs in ML, mainly focusing on DRL, and discuss more details about DRL's outstanding features and advantages to address future wireless networks' challenges.

Chapter 2: In the second chapter, we provide fundamental background and theory of the Markov decision process (MDP), a critical mathematical framework for modeling decision-making in situations where outcomes are partly random and partly under the control of a decision-maker. Specifically, essential components of an MDP and some typical extension models are presented. After that, specific solutions to address MDP problems, e.g. linear programming, value iteration, policy iteration, and reinforcement learning, are reviewed.

Chapter 3: Then, we will discuss DRL, a combination of RL and DL to address the current drawbacks of RL. In particular, we discuss more details how different DL models can be integrated into RL algorithms to speed up the learning processes. Many advanced DRL models are reviewed to provide a comprehensive perspective for the readers.

Chapter 4: In this chapter, we provide a particular scenario with detailed implementation to help the readers have a deeper understanding of step-by-step how to design, analyze, formulate, and solve an MDP optimization problem with DRL codes using conventional programming tools, e.g. TensorFlow. In addition, many simulation results are provided to discuss different aspects of implementing DRL and evaluate the impacts of parameters on learning processes.

Part II: Applications of DRL in Wireless Communications and Networking
This part focuses on studying diverse applications of DRL to address various problems in wireless networks, such as caching, offloading, resource sharing, and

security. We show example problems at the physical, media access control (MAC), network, and application layers and potential applications of DRL techniques to address them. Comparisons and detailed discussions are also provided to help the readers to have a comprehensive view of the advantages and limitations of using DRL to solve different problems in wireless networks.

Chapter 5 – DRL at the Physical Layer: The need for high-reliability and ultra-high capacity wireless communication has driven significant research into 5G communication systems. However, traditional techniques used for the design and optimization of these systems often struggle with the complexity and high dimensionality of the problems. In recent years, DRL has been recognized as a promising tool for addressing these complicated design and optimization problems. In this chapter, we will examine the potential applications of DRL in addressing three key issues at the physical layer of communication systems: beamforming, signal detection and channel estimation, and channel coding.

Chapter 6 – DRL at the MAC Layer: In modern networks like the IoT, sensors and mobile users often need to make independent decisions, such as selecting channels and base stations, in order to achieve their goals, such as maximizing throughput. However, this can be difficult due to the dynamic and uncertain nature of these networks. DRL algorithms like deep Q-learning (DQL) can help network entities learn and understand the states of the network, allowing them to make optimal decisions. In this chapter, we will review applications of DRL for addressing three issues at the MAC layer of these networks: resource allocation, channel access, and heterogeneous MAC protocols.

Chapter 7 – Network Layer: Network traffic control, which involves monitoring, inspecting, and regulating data flows, is a vital aspect of network systems that can improve information delivery efficiency and resource utilization. However, the proliferation of smart mobile devices in the IoT era and the future ultradense radio networks has greatly increased the scale of these networks and introduced highly dynamic topologies, leading to a rapid growth in data traffic and significant challenges in managing the Internet. In this chapter, we will discuss the potential applications of DRL to address three critical issues at the network layer: traffic routing, network slicing, and network intrusion detection.

Chapter 8 – Application and Service Layer: MEC has the potential to support computation-intensive applications, such as autonomous and cooperative driving, and reduce end-to-end latency in wireless and mobile networks. It also enables the development and deployment of intelligent systems and applications, such as intelligent transportation systems and smart city applications. However, there are several challenges that have yet to be addressed in the use of MEC, such as how to design and implement an efficient network framework, how to place MEC servers to serve a large number of users, and how to design offloading, caching, and communication strategies to improve system performance. In this chapter, we will review the applications of DRL frameworks in addressing three issues at the application and service layer: content caching, computation offloading, and data/computing task offloading.

Part III: Challenges, Approaches, Open Issues, and Emerging Research Topics

This part is dedicated to discussing some technical challenges along with open issues and introduces advanced approaches of DRL to address emerging issues in networking and communications.

Chapter 9: Although DRL has many advantages and has been developing rapidly, some technical limitations still need to be resolved. This chapter reviews some challenges in implementing DRL approaches in wireless networks, e.g. the curse of dimensionality, online learning, offline training processes, and quick adaptation to the uncertainty and dynamics of wireless environments. Several open issues in deploying DRL in wireless networks, e.g. inheriting and transferring knowledge among intelligent network entities, are discussed.

Chapter 10: In the last chapter, we introduce some emerging research topics in the context of applications of DRL in future wireless networks. These topics include joint radar and data communications, ambient backscatter communications, intelligent reflecting surface-aided communications, and rate splitting communications, as well as several advanced DRL models such as deep reinforcement transfer learning, generative adversarial networks, and meta-reinforcement learning.

Figure 1.12 presents the structure of the book.

1.4.2 Target Readership

The target readers for the proposed book consist of

- Researchers and communications engineers who are interested in studying applications of ML, especially RL and DRL for future wireless networks.
- Researchers and communications engineers who are interested in the state-of-the-art research on ML together with emerging problems for future autonomous communication systems.
- Developers and entrepreneurs who are interested in developing DRL applications in autonomous systems, e.g. smart homes, intelligent vehicles, and UAVs.
- Graduate students who are interested in learning and obtaining comprehensive knowledge on the design, analysis, and implementation of DRL to address practical issues in wireless networks.

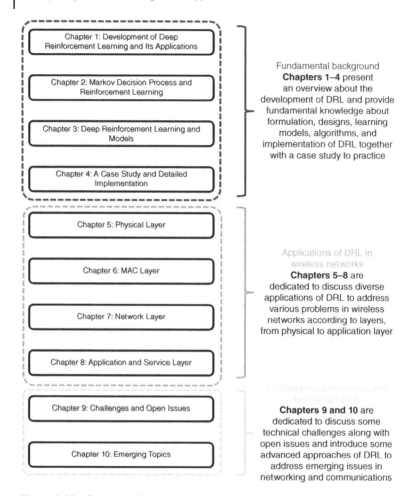

Figure 1.12 Structure of the book.

1.5 Chapter Summary

We have first discussed the development of wireless communications and the emerging challenges. Then, the fundamental of ML and DL as well as their development have been highlighted. After that, we have introduced some basic concepts of RL and DRL with their outstanding features in dealing with the emerging issues of future wireless communication systems. Finally, the structure of this book and target readership have been discussed.

References

1 Cisco, "Cisco annual internet report (2018–2023) white paper." https://www .cisco.com/c/en/us/solutions/collateral/executive-perspectives/annual-internet-report/white-paper-c11-741490.html, 2018. (Accessed on 10/01/2022).

2 X. Wang, Y. Han, C. Wang, Q. Zhao, X. Chen, and M. Chen, "In-edge AI: Intelligentizing mobile edge computing, caching and communication by federated learning," *IEEE Network*, vol. 33, no. 5, pp. 156–165, 2019.

3 Q.-V. Pham, F. Fang, V. N. Ha, M. J. Piran, M. Le, L. B. Le, W.-J. Hwang, and Z. Ding, "A survey of multi-access edge computing in 5g and beyond: Fundamentals, technology integration, and state-of-the-art," *IEEE Access*, vol. 8, pp. 116974–117017, 2020.

4 Y. Sun, M. Peng, Y. Zhou, Y. Huang, and S. Mao, "Application of machine learning in wireless networks: Key techniques and open issues," *IEEE Communications Surveys & Tutorials*, vol. 21, no. 4, pp. 3072–3108, 2019.

5 Y. C. Eldar, A. Goldsmith, D. Gündüz, and H. V. Poor (eds.), *Machine learning and wireless communications*, Cambridge University Press, Aug. 2022.

6 MATLAB, "Introducing deep learning with matlab." https://www.mathworks .com/campaigns/offers/next/deep-learning-ebook.html. (Accessed on 10/01/2022).

7 N. C. Luong, D. T. Hoang, S. Gong, D. Niyato, P. Wang, Y.-C. Liang, and D. I. Kim, "Applications of deep reinforcement learning in communications and networking: A survey," *IEEE Communications Surveys & Tutorials*, vol. 21, no. 4, pp. 3133–3174, 2019.

8 I. Goodfellow, Y. Bengio, and A. Courville, *Deep learning*. MIT Press, 2016.

9 C. Liu, J. Wang, X. Liu, and Y.-C. Liang, "Deep cm-cnn for spectrum sensing in cognitive radio," *IEEE Journal on Selected Areas in Communications*, vol. 37, no. 10, pp. 2306–2321, 2019.

10 J. Xie, J. Fang, C. Liu, and X. Li, "Deep learning-based spectrum sensing in cognitive radio: A CNN-LSTM approach," *IEEE Communications Letters*, vol. 24, no. 10, pp. 2196–2200, 2020.

11 D. Han, G. C. Sobabe, C. Zhang, X. Bai, Z. Wang, S. Liu, and B. Guo, "Spectrum sensing for cognitive radio based on convolution neural network," in *2017 10th international congress on image and signal processing, biomedical engineering and informatics (CISP-BMEI)*, pp. 1–6, IEEE, 2017.

12 A. P. Hermawan, R. R. Ginanjar, D.-S. Kim, and J.-M. Lee, "Cnn-based automatic modulation classification for beyond 5g communications," *IEEE Communications Letters*, vol. 24, no. 5, pp. 1038–1041, 2020.

13 S. Zheng, P. Qi, S. Chen, and X. Yang, "Fusion methods for cnn-based automatic modulation classification," *IEEE Access*, vol. 7, pp. 66496–66504, 2019.

14 H. Li, Y. Li, S. Zhou, and J. Wang, "Wireless channel feature extraction via gmm and cnn in the tomographic channel model," *Journal of communications and information networks*, vol. 2, no. 1, pp. 41–51, 2017.

15 D. Hong, Z. Zhang, and X. Xu, "Automatic modulation classification using recurrent neural networks," in *2017 3rd IEEE International Conference on Computer and Communications (ICCC)*, pp. 695–700, IEEE, 2017.

16 B. Bhargava, A. Deshmukh, M. V. Rupa, R. P. Sirigina, S. K. Vankayala, and A. Narasimhadhan, "Deep learning approach for wireless signal and modulation classification," in *2021 IEEE 94th Vehicular Technology Conference (VTC2021-Fall)*, pp. 1–6, IEEE, 2021.

17 N. Van Huynh, D. N. Nguyen, D. T. Hoang, T. X. Vu, E. Dutkiewicz, and S. Chatzinotas, "Defeating super-reactive jammers with deception strategy: Modeling, signal detection, and performance analysis," *arXiv preprint arXiv:2105.01308*, 2021.

18 Ö. Yildirim, "A novel wavelet sequence based on deep bidirectional LSTM network model for ecg signal classification," *Computers in biology and medicine*, vol. 96, pp. 189–202, 2018.

19 R. S. Sutton and A. G. Barto, *Reinforcement learning: An introduction*. MIT Press, 2018.

20 A. Feriani and E. Hossain, "Single and multi-agent deep reinforcement learning for AI-enabled wireless networks: A tutorial," *IEEE Communications Surveys & Tutorials*, 2021.

21 "Alphastar: Mastering the real-time strategy game starcraft ii."

22 "Muzero: Mastering go, chess, shogi and atari without rules."

23 "Automating pick and pack operations - the benefits."

24 I.-B. Park, J. Huh, J. Kim, and J. Park, "A reinforcement learning approach to robust scheduling of semiconductor manufacturing facilities," *IEEE Transactions on Automation Science and Engineering*, vol. 17, no. 3, pp. 1420–1431, 2019.

25 L. Zhou, L. Zhang, and B. K. Horn, "Deep reinforcement learning-based dynamic scheduling in smart manufacturing," *Procedia Cirp*, vol. 93, pp. 383–388, 2020.

26 K. Xia, C. Sacco, M. Kirkpatrick, C. Saidy, L. Nguyen, A. Kircaliali, and R. Harik, "A digital twin to train deep reinforcement learning agent for smart manufacturing plants: Environment, interfaces and intelligence," *Journal of Manufacturing Systems*, vol. 58, pp. 210–230, 2021.

27 P. M. Pilarski, M. R. Dawson, T. Degris, F. Fahimi, J. P. Carey, and R. S. Sutton, "Online human training of a myoelectric prosthesis controller via actor-critic reinforcement learning," in *2011 IEEE international conference on rehabilitation robotics*, pp. 1–7, IEEE, 2011.

28 C. Yu, J. Liu, S. Nemati, and G. Yin, "Reinforcement learning in healthcare: A survey," *ACM Computing Surveys (CSUR)*, vol. 55, no. 1, pp. 1–36, 2021.

29 A. Coronato, M. Naeem, G. De Pietro, and G. Paragliola, "Reinforcement learning for intelligent healthcare applications: A survey," *Artificial Intelligence in Medicine*, vol. 109, p. 101964, 2020.

30 M. Baucum, A. Khojandi, and R. Vasudevan, "Improving deep reinforcement learning with transitional variational autoencoders: A healthcare application," *IEEE Journal of Biomedical and Health Informatics*, vol. 25, no. 6, pp. 2273–2280, 2020.

31 "Kr102240442b1 - quality-aware deep reinforcement learning for proactive caching in millimeter-wave vehicular networks and system using the same."

32 "Cn109474980b - wireless network resource allocation method based on deep reinforcement learning."

33 "Cn110488861b - unmanned aerial vehicle track optimization method and device based on deep reinforcement learning and unmanned aerial vehicle."

34 D. Mwiti, "10 real-life applications of reinforcement learning," Nov 2021.

35 X. Di, K. Xiong, P. Fan, H.-C. Yang, and K. B. Letaief, "Optimal resource allocation in wireless powered communication networks with user cooperation," *IEEE Transactions on Wireless Communications*, vol. 16, no. 12, pp. 7936–7949, 2017.

36 M. Chu, H. Li, X. Liao, and S. Cui, "Reinforcement learning-based multiaccess control and battery prediction with energy harvesting in iot systems," *IEEE Internet of Things Journal*, vol. 6, no. 2, pp. 2009–2020, 2018.

37 J. Zhu, Y. Song, D. Jiang, and H. Song, "A new deep-q-learning-based transmission scheduling mechanism for the cognitive internet of things," *IEEE Internet of Things Journal*, vol. 5, no. 4, pp. 2375–2385, 2017.

2

Markov Decision Process and Reinforcement Learning

2.1 Markov Decision Process

In mathematics, a Markov decision process (MDP) is a discrete-time stochastic control process. Formally, an MDP describes a fully observable environment for reinforcement learning (RL). Theoretical results in RL are based on MDP, which means if the problem is formulated as an MDP, then the RL approach can be used to find solutions. Specifically, RL is a solution approach for MDP problems when the parameters of the MDP model are unknown. To understand MDP, we first present the *Markov property*, i.e. "the future is independent of the past given the present" [1, 2]. Once the current state is known, the information gathered from old experiences may not be necessary, and it is a sufficient statistic to predict the next action to take [2]. Mathematically, a state s_t has the Markov property, if and only if $P(s_{t+1}|s_t) = P(s_{t+1}|s_t, s_{t-1}, \dots, s_1)$.

An MDP models an environment in which all states have the Markov property. Formally, an MDP is tuple $(S, \mathcal{A}, P, r, \gamma)$, where

- S is the state space.
- $\mathcal{A}$ is the action space.
- P is a state transition probability function $P := S \times \mathcal{A} \mapsto [0,1]$.
- r is a reward function $r := S \times \mathcal{A} \times S \mapsto \mathbb{R}$.
- $\gamma \in [0,1]$ is the discount factor representing the trade-off between the immediate reward and upcoming rewards.

The full observability assumption allows the agent to have a full observation of the current state $s_t \in S$ in every time step t. Given the state s_t, the agent makes a *decision* of taking the action a_t and observes the new state s' in the MDP environment, where the s' is sampled from the transition probability $P(\cdot|s,a)$, and an immediate reward $r(s, a, s')$. Hence, the expected **return** is expressed as $\mathbb{E}\left[\sum_{t=0}^{\infty} \gamma^t r(s, a, s')|a \sim \pi(\cdot|s), s_0\right]$. This is also referred to as the *infinite-horizon discounted* return. Another popular formulation of the expected return named

Deep Reinforcement Learning for Wireless Communications and Networking:
Theory, Applications, and Implementation, First Edition.
Dinh Thai Hoang, Nguyen Van Huynh, Diep N. Nguyen, Ekram Hossain, and Dusit Niyato.
© 2023 The Institute of Electrical and Electronics Engineers, Inc. Published 2023 by John Wiley & Sons, Inc.

finite-horizon discounted return is expressed as $\mathbb{E}[\sum_{t=0}^{H} \gamma^t r(s, a, s') | a \sim \pi(\cdot|s), s_0]$, where the return is computed over a horizon H. This basically formulates the *episodic tasks* where the agent starts from an initial state and goes through a trajectory till encountering a terminal state. The principal goal of RL algorithm is learning a policy that maximizes the expected return.

Example: MDP formulations have been widely adopted to solve different control and/or decision problems in wireless networks. For instance, the authors in [3] formulated the anti-jamming problem in wireless networks as an MDP problem. The agent represents the legitimate transmitter that aims to transmit information to the receiver in the presence of a jammer. To defeat jamming attacks, the authors proposed to use the ambient backscatter communication technology to allow the transmitter to backscatter data right on the jamming signals. In the formulated MDP, the state consists of the number of packets in the data queue, the number of energy units in the energy queue, the state of the ambient radio frequency (RF) source (i.e. busy or idle), and the state of the jammer (i.e. attack or idle). Given a state, the agent (i.e. the legitimate transmitter) can choose to (i) stay idle, (ii) actively transmit data, (iii) harvest energy from the jamming signals and/or the ambient RF signals, (iv) backscatter data through the jamming signals and/or the ambient RF signals, and (v) adapt its transmission rate. The reward function is defined as the number of packets that are successfully transmitted to the receiver. The authors then used the deep Q-learning and deep dueling algorithms to obtain the optimal defense strategy for the transmitter. The simulation results demonstrated that deep reinforcement learning (DRL) can help to learn the jammer's strategy and effectively deal with the dynamics and uncertainty of the wireless environment, resulting in good communication performance under jamming attacks.

In a study by Wang et al. [4], the power allocation problem in an underwater full-duplex energy harvesting system was formulated as an MDP problem. The goal of the relay node (i.e. the "agent") is to maximize the long-term sum rate. At each step, the agent has access to information about the current battery level, the previous transmit power, and the channel state of the source-to-relay and relay-to-destination links. The agent selects the transmit power as its action and receives a reward equal to the data rate. The authors proposed a model-based algorithm to learn the optimal power allocation policy for this problem.

2.2 Partially Observable Markov Decision Process

In Section 2.1, we introduced MDP which defines the environment that the agent interacts with during the learning process. MDPs assume the *full* access to the state information. However, it is not always possible in real-world environments.

For instance, IoT devices collect measurements from their surrounding environment using sensors. These measurements are usually noisy and limited due to the quantizing process. Hence, the agent (e.g. centralized gateway) has only a *partial* access to the information about the real environment. A **partially observable Markov decision process** (POMDP) can be used to formulate environments, where the agent has limited/restrictive access to the states. POMDP is a 7-tuple $(S, A, P, r, \gamma, \mathcal{O}, Z)$ where the first five elements are the same as in MDP presented in 2.1 and the remaining elements are as follows:

- $\mathcal{O}$: the observation space.
- $Z := S \times A \times \mathcal{O} \mapsto [0,1]$: probability distribution over observations given state and action.

In particular, the observation space $\mathcal{O}$ consists of the information that the agent observes from the environment. Z presents the probability that the agent can obtain a particular observation after an action is taken in a state. Figure 2.1 illustrates and compares an MDP and a POMDP. Since with POMDP the agent does not have access to the states, we need to estimate the state of the environment. The two main approaches to solve the estimation issue are *history-based* and *predictive state representation* [6] methods. The **history-based** method estimates the current state using the data of the history of observation $H_t = \{o_1, \dots, o_{t-1}\}$ or the history of the observation-action pairs $H_t = \{(o_1, a_1), \dots, (o_{t-1}, a_{t-1})\}$. The history is exploited to learn the policy $\pi(\cdot|H_t)$ or the state-action values $Q(H_t, a_t)$ that we will discuss in-depth in Section 2.3. The use of all previous observations composes a large set of data to consider in the estimation of state; hence, we alleviate the history to a truncated version named to k-order Markov model. Nevertheless, the use of truncated history results in a loss of information that may be useful in the prediction of the state. As a result, we have a trade-off of how much of history we need to use because a long history requires more computation

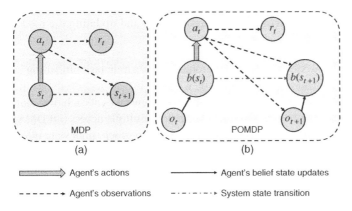

Figure 2.1 An illustration of (a) MDP and (b) POMDP. Luong et al. [5].

power and short history suffers from information loss. Also, it is noted that the choice of the value k is not straightforward [7] and requires tuning along with other hyper-parameters to improve the performance. Another method used to decrease the length of history over time of the state estimation process is the *belief state* $b(s) = p(s|H_t)$ that indicates the probability of being in state s. Here, the history is used to learn the probability distribution of being in state s rather than estimating the state s directly. The belief b is later used to learn the policy $\pi(\cdot|b)$ or the state-action value $Q(b, a)$ instead of the history. When the knowledge of $Z(s, a, o)$ is unknown, the belief states can be learned using the Bayes' rule [8].

The second method, named **predictive state representation** (PSR), makes use of future predictions instead of past actions and observations. The future is defined by m-length sequence of possible observations and actions $t = \{(o^1, a^1), \dots, (o^m, a^m)\}$, and also called in the literature by *test* sequence. The prediction of the future resides on computing the probability of the realization of the test observation after the history $h = \{(o_1, a_1), \dots, (o_m, a_m)\}$ have taken place. Mathematically, the probability distribution can be defined as follows:

$$P(t|h) = P(o_{n+1} = o^1, \dots, o_{n+m} = o^m | h, a_{n+1} = a^1, \dots, a_{n+m} = a^m). \quad (2.1)$$

Then, we build the *system-dynamics matrix* D by considering the history sequences as the rows and the test sequences as the columns. Thus, we get $D_{ij} = P(t_j|h_i)$. Supposing the $rank(D) = k < \infty$, there exists a set $C = \{c_1, \dots, c_k\}$ of k linearly independent columns referred to as the *core tests*. Hence, the future predictions t_j can be represented by a linear combination of the vectors in the set C for any history h_i. For a POMDP with a state space $|S| = N$, the system-dynamics matrix is with rank N [9]. It is noted that if we know the matrix D, we can get the model of the dynamics of environment. Therefore, the *linear PSR* model is expressed as follows [9]:

$$P(t|h) = P(C|h)m_t, \quad (2.2)$$

where m_t is the linear combination of the N vectors in C. In practice, learning a PSR model involves constructing the core tests $P(C|h)$ and updating the model parameters m_t [7].

Example: In a study by Xie et al. [10], the problem of dynamic task offloading in IoT fog systems was formulated as a POMDP, with the IoT device serving as the agent. The agent has a task queue that is updated with new tasks at each time slot, and the system uses an orthogonal frequency division multiple access (OFDMA) scheme with N subchannels. The agent must estimate the subchannel state information (CSI) and select a set of N_c sub-channels to contend for access. Because the agent may not be able to observe all information from all subchannels simultaneously, the problem is formulated as a POMDP. The fading process of each subchannel is modeled as a finite-state Markov chain (FSMC), with the CSI for a subchannel quantized into a finite number of discrete levels. The states in the model correspond to the quantized CSI and the queue state information, while the observations correspond to quantized channel states. Given an observation, the agent makes a decision on task offloading in order to minimize energy consumption while satisfying processing delay constraints. The authors proposed a history-based method to solve the POMDP, using a recurrent convolutional neural network to encode the action-observation history and learn the optimal Q-values.

2.3 Policy and Value Functions

The agent's behavior is determined by a **policy** that tells how the agent reacts at each state at every decision epoch. There are two types of policies: *deterministic* and *stochastic* policies, depending on the choice of the model. Deterministic policy is represented by a function $\pi(s_t) = a_t$ ($s_t \in S$ and $a_t \in A$) that maps directly the state space to the action space. Stochastic policy is a probability distribution over the action space, i.e. A: $\pi(a_t|s_t) = p_i \in [0,1]$, that returns the probability distribution of actions to execute given the current state. The policies are evaluated through the expected discounted return expressed as follows [1]:

$$R_t = \sum_{k=0}^{T} \gamma^k r_{t+k+1}, \tag{2.3}$$

where r_{t+k+1} is the immediate reward received at time step $t + k + 1$.

The discount factor γ introduced in the formulation of the MDP can be tuned in practice to enhance learning. Choosing a discount factor too small will lead to a greedy behavior of the agent, where it intends to learn to maximize the immediate reward. On the other hand, with a discount factor too close to 1, the future rewards will be slightly smaller than the immediate rewards. If the task is episodic (i.e. a terminal state exists), the discount factor can be set to 1, but if the task is continuous (i.e. the length of learning trajectories is infinite), the discount factor should be strictly less than 1. Theoretically, it can be proven that an optimal policy (denoted as π^*) exists, and it is as good or better than any other policy; however, it is hard to derive this best policy analytically. Therefore, the RL algorithms iteratively search for the optimal or near-optimal policies that maximize the expected return.

Value-function estimates *how good* a state is for the agent in terms of expected reward. A value function $V^\pi(s)$ is defined with respect to a policy π since the future rewards depend on actions that will be taken under that policy. Formally, $V^\pi(s)$ is defined as follows:

$$V^\pi(s) = E_\pi \left\{ R_t|s_t = s \right\} = E_\pi \left\{ \sum_{k=0}^{T} \gamma^k r_{t+k+1}|s_t = s \right\}, \tag{2.4}$$

where $E_\pi\{\}$ denotes the expected value given a policy π.

Similarly, we can define the value of taking an action a in state s with respect to policy π as follows:

$$Q^{\pi}(s, a) = E_{\pi} \left\{ R_t | s_t = s, a_t = a \right\} = E_{\pi} \left\{ \sum_{k=0}^{T} \gamma^k r_{t+k+1} | s_t = s, a_t = a \right\}, \quad (2.5)$$

where $Q^{\pi}(s, a)$ is the state-action value function for policy π. The value function can be expressed as sum of state-action values over all possible actions in the current state weighted by the probability of the action to be chosen (stochastic policy):

$$V^{\pi}(s) = \sum_{a \in A} \pi(s, a) \, Q^{\pi}(s, a). \quad (2.6)$$

The value functions V^{π} and Q^{π} are estimated from previous experiences sampled from the environment.

The problem of determining the optimal policy is referred to as *control problem* and determining value-function and action-state function is referred to as *prediction problem*. From optimal value function and optimal action-state value function, we can find the optimal policy as follows:

$$V^*(s) = \max_{\pi} \; V^{\pi}(s),$$

$$Q^*(s, a) = \max_{\pi} Q^{\pi}(s, a). \quad (2.7)$$

Therefore, the relationship between the optimal V^* and Q^* can be expressed as [1]:

$$V^*(s) = \max_{a \in A} Q^*(s, a). \quad (2.8)$$

2.4 Bellman Equations

A fundamental observation that is used in RL methods is that the value function can be defined recursively, e.g. we can write the state-value function as follows:

$$Q^{\pi}(s, a) = \sum_{s' \in S} P(s' | s, a) \left[r(s, a, s') + \gamma V^{\pi}(s') \right]. \quad (2.9)$$

Given that $R_t = r_{t+1} + \gamma R_{t+1}$, and putting it together with the equation relating the state value and state-action value Eq. (2.6), we can obtain the **Bellman** equations as follows:

$$V^{\pi}(s) = \sum_{s \in A} \pi(s, a) \sum_{s' \in S} P(s' | s, a)[r(s, a, s') + \gamma V^{\pi}(s')], \quad (2.10)$$

$$Q^{\pi}(s, a) = \sum_{s' \in S} P(s' | s, a) \left[r(s, a, s') + \gamma \sum_{a' \in A} \pi(s', a') Q^{\pi}(s', a') \right]. \quad (2.11)$$

The Bellman equations express the relationship between the state value/state-action value with its successor states and are used in dynamic programming (DP) methods to learn the value functions.

2.5 Solutions of MDP Problems

2.5.1 Dynamic Programming

DP is a class of methods to find exact solutions for a conventional MDP problem when the knowledge of the environment dynamic $P(\cdot|s, a)$ is known [1]. Basically, DP learns the value functions (*policy evaluation*) and uses them to search for good policy (*policy improvement*).

2.5.1.1 Policy Evaluation

It refers to computing the state-value function V^π for an arbitrary policy π. An iterative solution is adopted that uses a successive approximation obtained from the Bellman equation (Eq. (2.10)). Initially, we start from an arbitrary value of the state-value function V_0 (except terminal state, for which the given value should be 0), and we compute the next state-value function using the following update rule:

$$V_{k+1}(s) = \sum_a \pi(a|s) \sum_{s'} P(s'|s, a)[r(s, a, s') + \gamma V_k(s')], \qquad \forall s \in S. \qquad (2.12)$$

When $k \to \infty$, the convergence of the sequence $\{V_k\}$ is guaranteed under the same conditions that guarantee the existence of the optimal state-value function $V^*(s)$: either $\gamma < 1$ or eventual termination exists starting from any state [1]. The convergence condition is reached when there are no large changes in the state values: $\max_{s \in S} |V_{k+1}(s) - V_k(s)| < \epsilon$. This method is referred to as *iterative policy evaluation*.

2.5.1.2 Policy Improvement

It uses the policy evaluation process to find better policies starting from an arbitrary policy π_0. The policy improvement theorem [1] was introduced to compare policies and determine which policy is better than the others. Let π and π' denote any pair of deterministic policies such that for all states $s \in S$, $V^{\pi'}(s) \geq V^\pi(s)$. Then, the policy π' must be as good as, or better than, π. The new policy is defined by the greedy policy that maximizes the state-action value and constructed as follows:

$$\pi'(s) = \arg\max_a \sum_{s'} P(s'|s, a)[r(s, a, s') + \gamma V^\pi(s')]. \qquad (2.13)$$

2.5.1.3 Policy Iteration

The combination of the policy evaluation with policy improvement in a repeated computation results in **policy iteration**. Given a policy π_k, we perform policy evaluation to determine V^{π_k}. Then, by using policy improvement, we derive a new policy π_{k+1} which is better than the old policy.

Value iteration is an algorithm that learns the optimal state-value function from which the optimal policy is derived as follows:

$$\pi(s) = \arg\max_a \sum_{s'} P(s'|s,a)[r(s,a,s') + \gamma V^*(s')]. \tag{2.14}$$

It combines policy improvement with truncated policy evaluation steps into one update rule of the state-value function expressed as follows:

$$V_{k+1}(s) = \max_a \sum_{s'} P(s'|s,a)[r(s,a,s') + \gamma V_k(s')], \qquad \forall s \in S. \tag{2.15}$$

The update rule is repeated till convergence (i.e. $\max_{s \in S} |V_{k+1}(s) - V_k(s)| < \epsilon$).

2.5.2 Monte Carlo Sampling

A major limitation of DP methods is the requirement of the knowledge of the environment dynamics that may not be available in real-world problems. One way to solve this issue is estimating the state-value function $V(s)$ or the state-action value function $Q(s,a)$ based on samples of transitions from the environment rather than computing them using the transition probability function $P(\cdot|s,a)$. **Monte-Carlo sampling** uses this idea. It works as follows:

1. Start from state s_0.
2. Sample a trajectory at time t, $\tau_t = (s_0, a_0, r_1, s_1, a_2, r_3, s_3, \ldots, s_T, a_T, r_{T+1}, s_{T+1})$ from the environment using the current policy π.
3. Compute the discounted cumulative reward $R_t^{(e)} = \sum_{k=0}^{T} \gamma^k r_{t+k+1}$ for time step t.
4. Repeat the previous steps M times to obtain an estimate as follows:

$$V^\pi = \mathbb{E}_\pi [R_t|s_t = s] \approx \frac{1}{M} \sum_{t=1}^{M} R_t^{(e)}. \tag{2.16}$$

The values of $V^\pi(s)$ are approximated by averaging over multiple trajectories sampled from the environment; however, we cannot explicitly determine how many samples we need to have an accurate approximation. In practice, continuous updates of the estimated values are rather used:

$$V^\pi(s) \leftarrow V^\pi(s) + \alpha(R_t - V^\pi(s)), \tag{2.17}$$

where α is a weighting factor (a constant between 0 and 1). The state-action value function $Q(s,a)$ can be estimated using the same procedure:

$$Q^\pi(s,a) \leftarrow Q^\pi(s,a) + \alpha(R_t - Q^\pi(s,a)). \tag{2.18}$$

Monte Carlo sampling is generally used for episodic tasks where trajectories are finite. For continuous tasks, it is difficult to perform the averaging of rewards. Moreover, it heavily relies on exploitation so that the estimates converge to the optimal values. In fact, if we sample actions greedily from the current policy, the same actions will be chosen, and the agent will fall into a bad local minimum. To avoid the convergence to suboptimal solutions, we add randomness to the choice of the actions enforcing the exploration of new trajectories to cover a larger range of actions and trajectories in the estimation. However, in the case of fully random action selection, the evaluated policy does not correspond to the current policy anymore; nevertheless, it is rather a random policy. A compromise between these two approaches should be made for efficient learning. This issue is referred to as the *exploitation–exploration* dilemma. A practical way is to foster the random selection at the beginning of learning, and later on, rely more on exploiting the knowledge learned.

2.6 Reinforcement Learning

In Monte Carlo method, V^π is estimated using the average return in every episode. However, such technique slows down the training and is not feasible for infinite tasks. RL can be adopted to address this problem. Formally, RL is a computational approach to learning from interacting where an agent applies an action and receives a numerical signal, named reward, that informs the agent how good the action is. Rather than based on experts' experience to learn from, RL lets an agent learn on its own by focusing on maximizing the cumulative reward received from an environment. The outcome of an RL algorithm is a model that does the mapping between situations and actions to achieve a defined goal. RL is based on learning behaviors through interactions with an environment in discrete time steps. We can consider it like learning to play a new video game: an agent (gamer) tries to choose the best actions in the game (environment) to maximize the score (reward). Therefore, the interaction has two major steps, action, and perception. On every time step, the agent makes an observation on the environment o_t that describes the state s_t from state space $\mathcal{S}$. Then, according to a policy $\pi(a_t|s_t)$ ruling the behavior of the agent, the agent picks up an action a_t from the space $\mathcal{A}$ and applies it on the environment which leads to a new state s_{t+1}, and so on. The action choice is based on transition probability $P(s_{t+1}|s_t, a_t)$. Furthermore, the environment returns with the observation of a numerical signal (i.e. reward) $r(s, a)$ to give a feedback to the agent on the action a that has been chosen given the current state s. This process is repeated between the agent and the environment until a stop condition is satisfied.

In model-free RL, **temporal difference** (TD) is the most common method. In particular, this method replaces the average return with an estimate composed of the immediate reward sampled and an estimation of the expected return of the next state:

$$R_t = r(s, a, s') + \gamma R_{t+1} \approx r(s, a, s') + \gamma \underbrace{V^\pi(s')}_{\text{bootstrapping}} \ . \tag{2.19}$$

In TD methods, we can sample the immediate reward and estimate the return by $V^\pi(s')$; hence, it combines the sampling from Monte Carlo and bootstrapping from DP relying on the fact that returns can be expressed recursively. The update rules can be described as in the following equations:

$$V^\pi(s) \leftarrow V^\pi(s) + \alpha(\underbrace{r(s, a, s') + \gamma V^\pi(s') - V^\pi(s)}_{\delta}), \tag{2.20}$$

$$Q^\pi(s, a) \leftarrow Q^\pi(s, a) + \alpha(\underbrace{r(s, a, s') + \gamma Q^\pi(s', a') - Q^\pi(s, a)}_{\delta}). \tag{2.21}$$

Here, δ, which is named **TD error** or the **reward-prediction error**, defines the difference between the immediate reward summed with the return predicted in the next state/action and the current reward prediction. The main advantages of using TD methods are neither we need to wait for the end of an episode to update the state value function or the state-action value function, nor we need to have only episodic tasks. The main drawback is that the TD methods depend on an estimate that could be initially wrong. They require several iterations to start learning effectively.

When learning the Q-function, we distinguish the two ways for the selection of the next action a' in the update rule in Eq. (2.21), i.e. selecting the action using the policy $\pi(s)$, or using the greedy action $a^* = \arg\max_a Q(s', a)$. Two approaches are derived that are different in the choice of the next action a':

- **On-policy** method, e.g. **SARSA** (state-action-reward-state-action): This is a TD method, where the next action is sampled using the policy $\pi(s)$, and the TD error is given by Eq. (2.22). The policy should be ϵ-soft (e.g. ϵ-greedy or softmax) to ensure the exploration of different actions, i.e.

$$\delta = r(s, a, s') + \gamma Q^\pi(s', \pi(s')) - Q^\pi(s, a). \tag{2.22}$$

- **Off-policy** method, e.g. Q-learning: In this case, the greedy action (i.e. the action with the highest Q-value) is selected to update the current value as follows:

$$\delta = r(s, a, s') + \gamma \max_{a'} Q^\pi(s', a') - Q^\pi(s, a). \tag{2.23}$$

The algorithm for Q-learning is given in **Algorithm 2.1**. The selection of action a can be performed using the ϵ-greedy strategy as follows:

$$\text{action at iteration } k = \begin{cases} \arg\max Q(s, a) & \text{with probability } 1 - \epsilon, \\ \text{randomly} & \text{with probability } \epsilon. \end{cases} \tag{2.24}$$

On several benchmark environments, Q-learning has been shown to converge to the optimal policy even though the samples are not optimally collected. Note that the learning rate α needs to be decreased over time so that the Q-values do not hop

---

Algorithm 2.1 Tabular Q-learning

Initialize $Q_0(s, a) = 0 \; \forall s, a$.

Get initial state s_0.

while not converged, $k = 1, 2, \ldots$ **do**

    Choose action a, get next state s' and reward r.

    if s' is terminal **then**

        $y_{\text{target}} = r$.

        sample new initial state s'.

    else

        $y_{\text{target}} = r + \gamma \max_{a'} Q_k(s', a')$.

    end if

    $Q_{k+1}(s, a) \leftarrow (1 - \alpha)Q_k(s, a) + \alpha \, y_{\text{target}}$.

    $s \leftarrow s'$.

end while

---

around with each iteration, but at the same time, it should not be decreased too fast so that the new better actions can be learned. Additionally, it should satisfy the following conditions [11]:

$$\sum_{t=0}^{\infty} \alpha_t(s, a) = \infty, \qquad \sum_{t=0}^{\infty} \alpha_t^2(s, a) < \infty. \tag{2.25}$$

2.7 Chapter Summary

We have first presented the fundamentals of the MDP and the POMDP. Then, traditional solutions of the MDP such as dynamic programming, value iteration, and policy iteration are discussed in detail. Finally, reinforcement learning, the most well-known solution to the MDP, is reviewed.

References

1 R. S. Sutton and A. G. Barto, *Reinforcement learning: An introduction*. MIT Press, 2018.

2 D. Silver, "Lectures on reinforcement learning." URL: https://www.davidsilver .uk/teaching/, 2015.

3 N. Van Huynh, D. N. Nguyen, D. T. Hoang, and E. Dutkiewicz, "Jam me if you can": Defeating jammer with deep dueling neural network architecture and ambient backscattering augmented communications," *IEEE Journal on Selected Areas in Communications*, vol. 37, no. 11, pp. 2603–2620, 2019.

4 R. Wang, A. Yadav, E. A. Makled, O. A. Dobre, R. Zhao, and P. K. Varshney, "Optimal power allocation for full-duplex underwater relay networks with energy harvesting: A reinforcement learning approach," *IEEE Wireless Communications Letters*, vol. 9, no. 2, pp. 223–227, 2019.

5 N. C. Luong, D. T. Hoang, S. Gong, D. Niyato, P. Wang, Y.-C. Liang, and D. I. Kim, "Applications of deep reinforcement learning in communications and networking: A survey," *IEEE Communications Surveys & Tutorials*, vol. 21, no. 4, pp. 3133–3174, 2019.

6 M. Littman and R. S. Sutton, "Predictive representations of state," *Advances in neural information processing systems 14*, 2001.

7 A. Feriani and E. Hossain, "Single and multi-agent deep reinforcement learning for AI-enabled wireless networks: A tutorial," *IEEE Communications Surveys & Tutorials*, vol. 23, no. 2, pp. 1226–1252, 2021.

8 G. E. Monahan, "State of the art–a survey of partially observable Markov decision processes: Theory, models, and algorithms," *Management Science*, vol. 28, no. 1, pp. 1–16, 1982.

9 S. Singh, M. James, and M. Rudary, "Predictive state representations: A new theory for modeling dynamical systems," *arXiv preprint arXiv:1207.4167*, 2012.

10 R. Xie, Q. Tang, C. Liang, F. R. Yu, and T. Huang, "Dynamic computation offloading in IoT fog systems with imperfect channel-state information: A POMDP approach," *IEEE Internet of Things Journal*, vol. 8, no. 1, pp. 345–356, 2020.

11 T. Jaakkola, M. Jordan, and S. Singh, "Convergence of stochastic iterative dynamic programming algorithms," in *Advances in Neural Information Processing Systems* (J. Cowan, G. Tesauro, and J. Alspector, eds.), vol. 6, Morgan-Kaufmann, 1993.

3

Deep Reinforcement Learning Models and Techniques

3.1 Value-Based DRL Methods

Recall that the previous reinforcement learning (RL) methods in Chapter 2 fall under the class of *tabular methods* because the models $V^\pi(s)$ or $Q^\pi(s, a)$ are stored in a table containing the state values or the Q-values of every state-action pair. For these methods to converge, every pair of state-action should be encountered plenty of times. Tabular methods are only suitable for tasks with small state and action spaces. For tasks with large state and action spaces, the convergence time could be high. In many applications, most of the states may not be encountered, which is always the case with continuous states and actions. Therefore, it is more useful to generalize the learned Q-values: learn for the subset of state-action pairs and guess the newly encountered pairs based on the proximity between them. This is referred to as the *generalization* problem, which focuses on transferring the acquired knowledge to unseen data.

Function approximators can be effectively used in this problem as follows: instead of storing the Q-values in a table, represent them by a parameterized function $Q_\theta(s, a)$. This function approximates the optimal function $Q^*(s, a)$, which is optimized iteratively until it converges. We will focus on the neural networks as function approximators, but any other approximators (or regressors) such as linear models, radial-basis function networks, and support vector regression (SVR) should also work. When deep neural networks (DNNs) (Figure 1.2) are used as function approximators, we refer to the resulting RL models/methods as deep reinforcement learning (DRL) models/methods.

Value-based method is one of the two main categories of *model-free* RL methods along with the *policy-based methods*. In value-based models, an approximation of the value function $\hat{V}$ is learned, and the policy is obtained by acting *greedily* with respect to $\hat{V}$. In DRL, the function approximators are simply neural networks parameterized by θ, which are the weights for the connections between the layers. Generally, ϵ-greedy or softmax is adopted to select the actions, which

Deep Reinforcement Learning for Wireless Communications and Networking:
Theory, Applications, and Implementation, First Edition.
Dinh Thai Hoang, Nguyen Van Huynh, Diep N. Nguyen, Ekram Hossain, and Dusit Niyato.

make the policy to be depending directly on the Q-values generated by the function approximator $Q_\theta(s, a)$, and hence, on the weights θ. Henceforth, the policy is denoted by π_θ. Mathematically, the learning is defined by a minimization problem of a loss function which is defined based on the parameterized function $Q_\theta(s, a)$ and the real returns R_t sampled from the environment as follows:

$$\mathcal{L}(\theta) = \mathbb{E}_\pi[(R_t - Q_\theta(s, a))^2]. \tag{3.1}$$

For instance, by combining the Monte Carlo method with the function approximator, the expected return R_t is estimated through sampling of trajectories $\tau = (s_1, a_1, \ldots, a_{T-1}, a_{T-1}, a_T)$ from the environment dynamics. Consequently, the expectation is replaced by the *mean squared error* as follows:

$$\mathcal{L}(\theta) \approx \frac{1}{N} \sum_{i=1}^{N} \sum_{t=1}^{T} [R_t^{(i)} - Q_\theta(s, a)]^2, \tag{3.2}$$

where N is the number of trajectories and T is the length of the trajectories.

For SARSA and Q-Learning (see Section 2.6), the loss functions can be expressed as

$$\mathcal{L}(\theta) = \mathbb{E}_\pi\left[\left(\underbrace{r(s, a, s') + \gamma Q_\theta(s', \pi(s'))}_{\text{target}} - Q_\theta(s, a)\right)^2\right] \quad \text{SARSA,} \tag{3.3}$$

$$\mathcal{L}(\theta) = \mathbb{E}_\pi\left[\left(\underbrace{r(s, a, s') + \gamma \max_{a'} Q_\theta(s', a')}_{\text{target}} - Q_\theta(s, a)\right)^2\right] \quad \text{Q-learning.} \tag{3.4}$$

3.1.1 Deep Q-Network

Function approximators are useful in the case of tasks with large or continuous state and action spaces thanks to their generalization ability to accurately perform over unseen data. However, there are several issues when neural networks are used as function approximators for RL, especially when compared with supervised learning. First, the data in RL (i.e. transitions) can be highly correlated which can affect the training of a neural network and result in a poor local minimum. Also, the target values in the loss function (Eqs. (3.3) and (3.4)) depend on the function we are trying to learn. On the other hand, in a supervised learning setup, e.g. in classification and regression problems, the target values, namely, the labels, are constants during the learning, and we optimize the model to fit to the labels. Meanwhile, in RL update rules, for instance, the

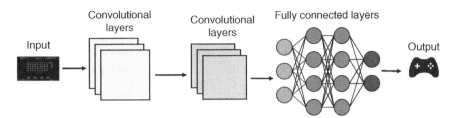

Convolutional layers Convolutional layers Fully connected layers

Input Output

Figure 3.1 Classical DQN architecture predicts Q-values. Adapted from Mnih et al. [1].

target value $r(s, a, s') + \gamma\max_{a'} Q_\pi(s', a')$ in Q-learning varies during the learning process because it depends on the approximator we are optimizing. As a result, the learning suffers from the instability problem, and it requires a large amount of updates for the function approximator (i.e. the neural network) to converge.

Deep Q-network (DQN), proposed by Mnih et al. [1], is an effective implementation of function approximators in RL methods to overcome the issues related to correlated inputs/outputs and nonstationarity in RL, as illustrated in Figure 3.1. To deal with the correlated inputs/outputs, the idea proposed here is quite simple: instead of updating the Q-function successively with each transition, store the experienced transitions in an **experience replay memory** or **replay buffer** with a large capacity (e.g. 10^6 transitions). Then, the stochastic gradient descent (SGD) method samples randomly from the experience replay memory forming minibatches and train the neural network by minimizing the loss function while hoping the samples are uncorrelated. The new transitions replace the old ones if the buffer is entirely filled. Moreover, DQN improves the stability of the learning of Q-function and solves the nonstationarity aspect of the target values $r(s, a, s') + \gamma\max_{a'} Q_\theta(s', a')$ by computing them with a **target network** noted $Q_{\theta'}$ that is identical to the actual Q-function Q_θ. The target network is updated less frequently (e.g. every thousands of updates) with the most recent version of the learned parameters θ. Hence, the target values remain constant for a sufficient number of iterations, and they become more stationary. The DQN method is summarized in **Algorithm 3.1**.

Note in **Algorithm 3.1**, we introduce a general loss function $\mathcal{L} : \mathbb{R} \to \mathbb{R}^+$. In Q-learning, we generally use the squared loss, but we can use other loss functions as well. For example, Huber loss illustrated in Figure 3.2 can be used to prevent overfitting on the targets because it is known for its robustness in presence of outliers. Huber loss is defined as follows (as illustrated in Figure 3.2):

$$L_\delta(x) = \begin{cases} \frac{1}{2}x^2 & \text{for } |x| < \delta, \\ \delta(|x| - \frac{1}{2}\delta) & \text{otherwise.} \end{cases} \tag{3.6}$$

Moreover, various optimization algorithms can be used to minimize the loss function $\mathcal{L}(\theta)$ such as Adam, RMSProp, or vanilla SGD.

Algorithm 3.1 Deep Q-network (DQN) algorithm with experience replay

---

1: Initialize replay buffer B to store the transitions.
2: Initialize randomly the Q-function Q_θ and the target network $Q_{\theta'}$ such that $\theta = \theta'$.
3: Observe initial state s_1.
4: **for** $t = 1, \dots, T$ **do**
5:     Select action a_t based on the behavior policy derived from $Q_\theta(s_t, a_t)$ (e.g. ϵ-greedy).
6:     Observe the new state s_{t+1} and reward r_t.
7:     Store transition (s_t, a_t, r_t, s_{t+1}) in the replay memory B.
8:     Sample minibatch of transitions (s_j, a_j, r_j, s_{j+1}) from B.
9:     Set

$$y_j = \begin{cases} r_j & \text{if } s_{j+1} \text{ is terminal,} \\ r_j + \gamma \max_{a'} Q_{\theta'}(s_{j+1}, a') & \text{otherwise.} \end{cases} \quad (3.5)$$

10:     Optimize $\mathcal{L}(\theta) = \mathcal{L}\left(Q_\theta(s_j, a_j) - y_j\right)$ using SGD (or its variants).
11:     Every C steps reset $\theta' = \theta$.
12: **end for**

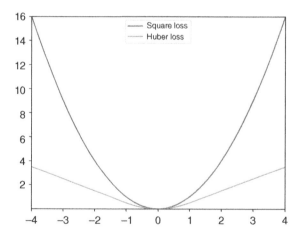

Figure 3.2 Huber loss and square loss.

We emphasize that the transitions stored in the replay buffer are sampled uniformly randomly for training. Therefore, the recent transitions may be used for training in later time than the time experienced at. Meanwhile, old transitions that are generated using bad policy can still be used for training multiple times.

With DQN, the target network is updated less often than the learned network, so the target values will be wrong at the beginning of the learning. Wrong targets lead the actual Q-function to converge to a suboptimal solution that results in a bad policy. More recent algorithms such as DDPG (see Section 3.3.2 for more details) that use the same concept to learn the Q-function (also known as the critic network), use *polyak* updates for the target network as follows:

$$\theta' = \tau\theta + (1-\tau)\theta', \tag{3.7}$$

where $\tau \ll 1$ is the polyak parameter. Polyak update is a smooth update of the weights where the target network steadily tracks the actual Q-function but never will be the same.

It is well known that DQN algorithm learns slowly and it requires a large number of samples to achieve a satisfactory policy. In DRL, this is referred to as the **sampling complexity**, i.e. the amount of samples from the environment required by an algorithm to obtain satisfactory performance.

Example: In the wireless communications, DQN is the go-to RL algorithm for solving problems with discrete action spaces. For instance, in [2], the authors used DQN to solve the dynamic multichannel access problem in the wireless networks. Simulation results demonstrated that, by using DRL, the performance of the proposed solution is much better than conventional algorithms. For problems with continuous action spaces, researchers use discretization schemes to fit the problem to a Q-learning framework [3].

3.1.2 Double DQN

Double DQN was proposed by Van Hasselt et al. [4] as an improvement of DQN. As a matter of fact, DQN suffers from an upward bias in the target value $r(s, a, s') + \gamma Q_{\theta'}(s, a)$. The target value can be overestimated because of the max operator so the expected return would be far higher than the true values. This happens especially in the first iterations when $Q_{\theta'}(s', a)$ is high than the true value. This may lead to an overestimated action because the next action is the action with the highest Q-value returned by the target network $Q_{\theta'}$. This will affect the learned Q-values $Q_\theta(s, a)$ that will also be overestimated. In [5], **double learning** was suggested to overcome the overestimation in regular Q-learning.

To overcome this overestimation, we select the next action with a greedy policy based on the learned Q-function Q_θ and use the target network $Q_{\theta'}$ to estimate the expected return rather than selecting the action with a greedy policy based on the target network $Q_{\theta'}$. The expression for the target value then turns into

$$y = r(s, a, s') + \gamma Q_{\theta'}\left(s', \arg\max_{a'} Q_\theta(s', a')\right). \tag{3.8}$$

This implies a small modification in the DQN algorithm, but adds a significant improvement to its performance and stability.

Example: [6] used deep double Q-network (DDQN) jointly with a Deep Deterministic Policy Gradient (DDPG) method which will be discussed later in Section 3.3.2, to perform dynamic clustering of access points (APs) in cell-free networks for the beamforming optimization problem. The joint clustering and beamforming optimization problem is formulated as an Markov decision process (MDP) problem, and the training is performed in a central unit within the cell-free network.

3.1.3 Prioritized Experience Replay

Another idea often used along with the replay buffer is the **prioritized experience replay** (PER) [7]. As a matter of fact, replaying all transitions with equal probability is highly suboptimal. Some transitions are more interesting and valuable than others, thus the uniform sampling of old transitions slows down the learning. The importance of each transition can be measured by the temporal difference (TD)-error δ, which indicates how *surprising* the transition is compared to the expected value estimated:

$$\delta = r(s, a, s') + \gamma Q_{\theta'}\left(s', \arg\max_{a'} Q_\theta(s', a')\right) - Q_\theta(s, a) \quad \text{DDQN}. \tag{3.9}$$

One intuitive idea is *greedy TD-error prioritization*, where the transitions are stored with their TD-errors $(s_j, a_j, r_j, s_{j+1}, \delta_j)$, and later on, the transitions that have the highest TD-error are sampled from the replay buffer. The Q-learning update rule is applied to these transitions where the gradient is weighted by the TD-error. This method shows an improvement compared to uniform sampling; however, it still has several issues. The transitions with low TD-error on first visits may never be sampled during the training. Also, the greedy prioritization suffers from the lack of diversity. It may lead to overfitting on the subset of samples that have initially high TD-errors since the errors shrink slowly. Therefore, the stochastic sampling is more suitable to overcome these issues. The key idea is to give the transitions monotonic nonzero probabilities:

$$P(i) = \frac{p_i^\alpha}{\sum_k p_k^\alpha}, \tag{3.10}$$

where $p_i > 0$ is the transition probability and α determines how much prioritization is used ($\alpha = 0$ corresponds to uniform sampling).

The choice of p_i can be done in two ways:

- $p_i = |\delta_i| + \epsilon$, where epsilon is used to prevent the extreme cases of transitions with zero errors from not being visited.
- $p_i = \frac{1}{\text{rank}(i)}$ where the rank is defined by the ordering of transitions according to $|\delta_i|$.

The second distribution is insensitive to outliers, which make it more likely to be robust than the first distribution. Additionally, PER introduces bias in the estimation of the expected value. The expectation $\mathbb{E}_\pi$ in Q-learning (Eq. (3.4)) is based on uniform sampling; however, with prioritization, we are using other sampling distribution that leads to a different expectation. This bias is corrected using *importance sampling* (IS) weights to compensate for the nonuniform probabilities $P(i)$ as follows:

$$w_i = \left(\frac{1}{N} \frac{1}{P(i)} \right)^\beta. \tag{3.11}$$

More details on this can be found in the original paper [7].

To make PER more practical, DeepMind uses special data structures to store the transitions in order to reduce the time complexity of certain operations because of the large amount of data that we need to train the neural networks. In OpenAI implementations, the sum-tree data structure was used for the replay buffer which makes both updating and sampling operations possible with a complexity of $O(\log N)$. The training steps using PER are summarized in **Algorithm 3.2**.

Algorithm 3.2 DDQN with prioritized experience replay [7]

1: **Input:** size of minibatch k, learning rate η, update period K, replay size N, exponents α and β, and total number of iterations T.
2: Initialize replay memory $\mathcal{B} = \emptyset$, $\Delta = 0$, $p_0 = 1$.
3: Observe state s_1.
4: **for** $t = 1, \ldots, T$ **do**
5:     Execute action $a_t \sim \pi_\theta(s_t)$.
6:     Observe new state s_{t+1} and receive reward r_t.
7:     Store transition (s_t, a_t, r_t, s_t) in $\mathcal{B}$ with maximal priority $p_t = \max_{i<t} p_i$.
8:     **if** $t \equiv 0 \mod K$ **then**
9:         **for** $j = 1, \ldots, k$ **do**
10:             Sample transition $j \sim P(i) = p_j^\alpha / \sum_i p_i^\alpha$
11:             Compute IS weight $w_j = (N \cdot P(j))^\beta / \max_i w_i$.
12:             Compute TD-error $\delta_j = r_j + \gamma Q_{\theta'}(s_{j+1}, \arg\max_a Q_\theta(s_{j+1}, a)) - Q_\theta(s_j, a_j)$.
13:             Update transition priority $p_j \leftarrow |\delta_j|$
14:             Accumulate weight-change $\Delta \leftarrow \Delta + w_j.\delta_j \nabla_\theta Q_\theta(s_j, a_j)$
15:         **end for**
16:         Update weights $\theta \leftarrow \theta + \eta \cdot \Delta$ and reset $\Delta = 0$.
17:         From time to time copy weights into target network $\theta' \leftarrow \theta$.
18:     **end if**
19: **end for**

3.1.4 Dueling Network

The traditional DQN architecture employs a single neural network to directly estimate the Q-values of all possible actions $Q_\theta(s, a)$. However, Baird in [8] motivated the decomposition of the Q-values as a summation of the state-value and the *advantage* of selecting an action:

$$Q^\pi(s, a) = V^\pi(s) + A^\pi(s, a). \tag{3.12}$$

The action value is determined by (i) the value of the underlying state s, $V^\pi(s)$, and (ii) the interest of choosing that action $A^\pi(s, a)$. A zero advantage corresponds to choosing an optimal action that maximizes the expected return (i.e. equal to the state-value). Also, the advantage of the other actions is negative which makes them less advantageous. Moreover, the range of the advantages is narrower than Q-values as the latest ones are theoretically not bounded. Thus, it is more convenient for the neural networks to learn the advantages $A_\theta(s, a)$ compared to learning the Q-function $Q_\theta(s, a)$.

In [9], the authors proposed a novel way to incorporate the advantage function in double DQN architecture with prioritized replay as shown in Figure 3.3. The output layer is the same for both architectures, and it predicts the Q-values with the same loss function:

$$\mathcal{L}(\theta) = \mathbb{E}_\pi\left[\left(r(s, a, s') + \gamma Q_{\theta', \alpha', \beta'}\left(s', \arg\max_{a'} Q_{\theta, \alpha, \beta}(s', a')\right) - Q_{\theta, \alpha, \beta}(s, a)\right)^2\right].$$

$$\tag{3.13}$$

The difference is that the layer before the output layer predicts separately the advantage of actions $A_{\theta, \alpha}(s, a)$ and state-value $V_{\theta, \beta}(s)$. θ represents the weights of

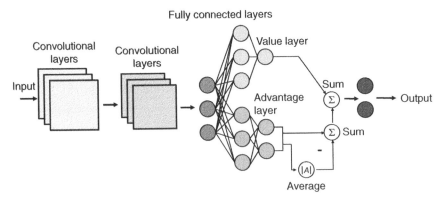

Figure 3.3 Dueling network predicting separately the advantages and state values to compute the Q-values. Adapted from Wang et al. [9].

the early layers, and α and β are the weights of the separated networks to predict $A_{\theta,\alpha}$, and $V_{\theta,\beta}$. The Q-value is formed by an addition of the two subnetworks:

$$Q_{\theta,\alpha,\beta}(s,a) = V_{\theta,\beta}(s) + A_{\theta,\alpha}(s,a). \tag{3.14}$$

One issue with Eq. (3.14) is that V and Q cannot be recovered uniquely, e.g. adding a constant to A and subtracting it from V leads to the same Q-value. Therefore, we normalize the advantages to ensure that the greedy action has a zero advantage:

$$Q_{\theta,\alpha,\beta}(s,a) = V_{\theta,\beta}(s) + \left(A_{\theta,\alpha}(s,a) - \max_a A_{\theta,\alpha}(s,a) \right). \tag{3.15}$$

It was reported in [9] that it is more efficient to substitute the max operator with the mean of the advantages. Hence, in place of making quick compensation to the changes of the greedy actions along with the improvement of the policy, the advantages would simply require to follow the mean changes velocity:

$$Q_{\theta,\alpha,\beta} = V_{\theta,\beta}(s) + \left(A_{\theta,\alpha}(s,a) - \frac{1}{|\mathcal{A}|} \sum_a A_{\theta,\alpha}(s,a) \right). \tag{3.16}$$

Finally, everything in the training process apart from the architecture of the output layer remains the same. Specifically, the gradient of the mean square error (MSE) loss function can be generated through backpropagation to update the weights.

Example: Dueling networks are considered as an amelioration of DQN networks. In [10], the dueling network combined with DDQN was adopted to solve the user association and resource allocation optimization problem in heterogeneous cellular networks. The authors developed a multiagent reinforcement learning (MARL)-based solution, namely, D3QN, using the dueling double DQN performing a distributed optimization method. The simulations showed that D3QN achieved better performance than other RL methods and converged rapidly to the subgame perfect Nash equilibrium.

3.2 Policy-Gradient Methods

In this section, we discuss the policy search methods in DRL that focus on learning directly the policy π_θ, which is generally represented by a neural network with weights θ. The learning goal is to maximize an objective function that represents the return over a trajectory. Mathematically, the objective function is defined as follows:

$$J(\theta) = \mathbb{E}_{\tau \sim p_\theta}[R(\tau)] = \mathbb{E}_{\tau \sim p_\theta}\left[\sum_{t=0}^{T} \gamma^t r(s_t, a_t, s_{t+1}) \right], \tag{3.17}$$

where $\tau = (s_0, a_0, s_1, a_1, \ldots, s_T, a_T)$ is the trajectory and p_θ is the distribution of trajectories. The likelihood of sampling a trajectory τ can be computed as follows:

$$p_\theta(\tau) = p_0(s_0) \prod_{t=0}^{T} \pi_\theta(a_t, s_t) p(s_{t+1} | s_t, a_t), \tag{3.18}$$

where $p_0(s_0)$ defines the probability of starting from state s_0 and $p(s_{t+1} | s_t, a_t)$ is the transition probability defining the MDP. However, we do not want to use this term to define the objective because we do not acquire the knowledge of the system dynamics. Instead, we can estimate the objective function $J(\theta)$ using trajectories sampled from the trajectory distribution p_θ as shown in Eq. (3.17). Using Monte Carlo method, we approximate the objective as follows:

$$J(\theta) = \mathbb{E}_{\tau \sim p_\theta(\tau)} \left[\sum_t r(s_t, a_t) \right] \approx \frac{1}{N} \sum_i \sum_t r(s_t, a_t). \tag{3.19}$$

The approximation is based on the generation of multiple trajectories $\{\tau_i\}$ by running the policy and using the average of trajectory rewards. Consequently, the more the number of trajectories generated, the higher accuracy the evaluation objective can achieve.

To learn the policy that maximizes the objective function $J(\theta)$, we apply gradient ascent on the weights θ. Therefore, we require the gradient of the objective function w.r.t. the weights:

$$\nabla_\theta J(\theta) = \frac{\partial J(\theta)}{\partial \theta}. \tag{3.20}$$

Then, the update rule for the weights can be given as follows:

$$\theta \leftarrow \theta + \eta \nabla_\theta J(\theta). \tag{3.21}$$

3.2.1 REINFORCE Algorithm

The **REINFORCE** algorithm was proposed in 1992 by Williams [11] that can estimate the policy gradient using Monte Carlo method. In this part, we will introduce the mathematical derivations that lead to REINFORCE algorithm and discuss techniques used to reduce the variance of estimating the policy gradient.

3.2.1.1 Policy Gradient Estimation

Denoting the trajectory reward by $R(\tau) = \sum_{t=1}^{T} r(s_t, a_t)$, the objective can be expressed as follows:

$$\begin{aligned} J(\theta) &= \mathbb{E}_{\tau \sim p_\theta(\tau)}[R(\tau)] \\ &= \int p_\theta(\tau) R(\tau) d\tau. \end{aligned} \tag{3.22}$$

Therefore, the policy gradient can be written as

$$\nabla_\theta J(\theta) = \int \nabla_\theta p_\theta(\tau) R(\tau) d\tau. \tag{3.23}$$

We apply the *log-trick* on $\nabla_\theta p_\theta(\tau)$ which is expressed as follows:

$$f(x)\nabla_\theta \log f(x) = f(x)\frac{\nabla_\theta f(x)}{f(x)} = \nabla_\theta f(x). \tag{3.24}$$

Therefore, the policy gradient can be given by

$$\begin{aligned} \nabla_\theta J(\theta) &= \int p_\theta(\tau)\nabla_\theta \log p_\theta(\tau) R(\tau) d\tau \\ &= \mathbb{E}_{\tau \sim p_\theta(\tau)} \left[\nabla_\theta \log p_\theta(\tau) R(\tau) \right]. \end{aligned} \tag{3.25}$$

The expectation can be evaluated by using samples from the environment. Moreover, by expressing $\log p_\theta(\tau)$ based on the expression in Eq. (3.18), we have

$$\nabla_\theta \log p_\theta(\tau) = \sum_{t=1}^{T} \log \pi_\theta(a_t|s_t). \tag{3.26}$$

Therefore, the expression for the gradient can be written in the following form:

$$\nabla_\theta J(\theta) = \mathbb{E}_{\tau \sim p_\theta(\tau)} \left[\left(\sum_{t=1}^{T} \nabla_\theta \log \pi_\theta(a_t|s_t) \right) \left(\sum_{t=1}^{T} r(s_t, a_t) \right) \right]. \tag{3.27}$$

To evaluate the policy gradient, we use Monte Carlo method by sampling from the trajectory distribution:

$$\nabla_\theta J(\theta) \approx \frac{1}{N} \sum_{i=1}^{N} \left(\sum_{t=1}^{T} \nabla_\theta \log \pi_\theta(a_{i,t}|s_{i,t}) \right) \left(\sum_{t=1}^{T} r(s_{i,t}, a_{i,t}) \right), \tag{3.28}$$

where $s_{i,t}$ and $a_{i,t}$ are the tth state and action in the ith trajectory, respectively. Then, we update the weights of the neural network of the policy using the gradient descent algorithm as follows:

$$\theta \leftarrow \theta + \alpha \nabla_\theta J(\theta). \tag{3.29}$$

The basic REINFORCE algorithm can be summarized in the following three main steps:

i. *Trajectory sampling*: Run the policy $\pi_\theta(\cdot)$ for multiple trajectories $\{\tau_i\}$.
ii. *Estimation of the policy gradient*: Evaluate the policy gradient $\nabla_\theta J(\theta)$ using Monte Carlo method (Eq. (3.28)).
iii. *Improve the policy*: Compute the new weights θ using gradient ascent (Eq. (3.29)).

3.2.1.2 Reducing the Variance

The policy gradient methods suffer from sample inefficiency in terms of computation of the gradient $\nabla_\theta J(\theta)$. In Eq. (3.28), we estimate the gradient using Monte Carlo by running multiple trajectories, and in order to get a good estimate of the gradient, we need to sample a large number of trajectories, which is expensive. Additionally, for each trajectory, we sum over rewards received in each step, which incurs a large computational cost to obtain a good estimate of the gradient. The policy gradient methods also suffers from slow convergence because we need a large number of samples to update the policy.

In addition to the computational difficulties, the policy gradient suffers from high variance. To see this problem, we consider large problems as learning the resource allocation in an unknown environment or training an agent to predict the beamforming elements in large networks. While sampling trajectories from an untrained policy, we are more likely to go across largely different behaviors. For an effective learning of the optimal policy, the agent requires to observe different actions in different states which are too large in real environments, and it grows exponentially in continuous state and action spaces because visiting every action-state pair becomes computationally intractable. Using Monte Carlo estimator, we are trading off between the computational feasibility and the gradient accuracy.

Baseline One thing to notice in the policy gradient descent is the heavy reliability on the trajectory reward $R(\tau) = \sum_t r(a_t, s_t)$. If the trajectory reward $R(\tau_1)$ is negative, then the policy π_θ goes one step in the opposite direction of the gradient which should lessen the probability on that trajectory $\pi_\theta(\tau_1) < 0$. On the other hand, if the trajectory reward is positive, i.e. $R(\tau_2) > 0$, the probability density $\pi_\theta(\tau_2)$ will increase so that the policy takes those trajectories more often. If we add a constant to the trajectory rewards $R(\tau) + b$ so that $R(\tau_1) + b > 0$, then the probability weight on τ_1 will increase even though it is worse than other trajectories. Thus, the policy gradient is too sensitive to the shifting and scaling of the reward function, and from this point, we want to find an optimal b that minimizes the variance of the policy gradient. Taking into account the **baseline** b, the policy gradient expression becomes

$$\nabla_\theta J(\theta) \approx \frac{1}{N} \sum_{i=1}^{N} \nabla \log \pi_\theta(\tau_i) \left(R(\tau_i) - b \right). \tag{3.30}$$

One natural definition of b could be the average rewards on all the trajectories $\frac{1}{N} \sum_{i=1}^{N} R(\tau_i)$, so that the probabilities of trajectories that have better rewards would be increased and the probabilities for the bad trajectories will be decreased and therefore less preferred. However, two main questions arising at this point (i) are we allowed to formulate the policy gradient as in Eq. (3.30)? (ii) if yes, what is the optimal value of b to reduce the variance? Responding to the first question, we need to verify if the estimate of the policy gradient is unbiased which is equivalent to showing that $\mathbb{E}_{\tau \sim p_\theta(\tau)}[\nabla_\theta \log \pi_\theta(\tau)b]$ is equal to zero. We have

$$
\begin{aligned}
\mathbb{E}_{\tau \sim p_\theta(\tau)} \left[\nabla_\theta \log \pi_\theta(\tau)b \right] &= \int b\, \pi_\theta(\tau) \nabla_\theta \log \pi_\theta(\tau) \mathrm{d}\tau \\
&= b \int \nabla_\theta \pi_\theta(\tau) \mathrm{d}\tau \qquad (3.31) \\
&= b \nabla_\theta 1 \\
&= 0.
\end{aligned}
$$

Therefore, the policy gradient with a baseline remains an unbiased estimator. Next, to derive the expression of the optimal baseline b^*, we express the policy gradient's variance as follows:

$$
\begin{aligned}
\mathrm{Var}[\nabla_\theta J(\theta)] =\ & \mathbb{E}_{\tau \sim p_\theta(\tau)} \left[\left(\nabla_\theta \log \pi_\theta(\tau)(R(\tau) - b) \right)^2 \right] \\
& - \mathbb{E}_{\tau \sim p_\theta(\tau)} \left[\nabla_\theta \log \pi_\theta(\tau)(R(\tau) - b) \right]^2.
\end{aligned}
\qquad (3.32)
$$

For simplicity, let us introduce the term $g(\tau) = \nabla_\theta \log \pi_\theta(\tau)$. Since the baselines are unbiased in expectation, the second term of the variance is simply equal to the policy gradient $\mathbb{E}[\nabla_\theta \log \pi_\theta(\tau)R(\tau)]$, which is independent of b. Therefore, we focus on the first term, and we calculate the first derivative of the variance with respect to b:

$$
\begin{aligned}
\frac{\mathrm{dVar}}{\mathrm{d}b} &= \frac{\mathrm{d}}{\mathrm{d}b} \mathbb{E} \left[g(\tau)^2 (R(\tau) - b)^2 \right] \\
&= \frac{\mathrm{d}}{\mathrm{d}b} \left(\mathbb{E} \left[g(\tau)^2 R(\tau)^2 \right] - 2\mathbb{E} \left[g(\tau)^2 R(\tau) b \right] + b^2 \mathbb{E} \left[g(\tau)^2 \right] \right) \qquad (3.33) \\
&= 0 - 2\mathbb{E} \left[g(\tau)^2 R(\tau) \right] + 2b\, \mathbb{E} \left[g(\tau)^2 \right].
\end{aligned}
$$

The first derivative of the variance is equal to zero when

$$
b^* = \frac{\mathbb{E}[g(\tau)^2 R(\tau)]}{g(\tau)^2}.
\qquad (3.34)
$$

In other words, the optimal baseline to consider is to re-weight the expected returns (i.e. trajectory reward) by its expected policy gradient magnitude $g(\tau)^2$. However, this optimal baseline is hard to implement in RL algorithms. In practice, one popular way is to define the baseline as the value function $V^\pi(s_t)$ that determines the expected return while being in state s_t and behaving under the current policy π:

$$
b(s_t) = \mathbb{E}[r_t + \gamma r_{t+1} + \ldots + \gamma^T r_{t+T}] = V^\pi(s_t).
\qquad (3.35)
$$

By using the state values $V^\pi(s_t)$, the log probability of action increases proportionally to how much its returns are better than the expected return under the

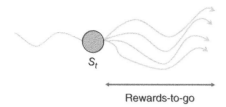

Figure 3.4 Reward-to-go: summation of rewards collected starting from transition (s_t, a_t).

s_t

Rewards-to-go

current policy. However, the value function still needs to be learned, which makes the algorithm more complex.

Causality To further minimize the variance of estimation of the policy gradient in REINFORCE algorithm, we make use of the universal principle of *causality*: rewards in the present, at time t, can only influence rewards in that time and beyond $t' \geq t$. Therefore, Eq. (3.28) for the policy gradient becomes

$$\nabla_\theta J(\theta) \approx \frac{1}{N} \sum_{i=1}^{N} \sum_{t=1}^{T} \nabla_\theta \log \pi_\theta(a_{i,t}|s_{i,t}) \left(\sum_{t'=t}^{T} r(s_{i,t'}, a_{i,t'}) \right). \tag{3.36}$$

The log-likelihood, also called the score function, $\nabla_\theta \log \pi_\theta(a_{i,t}|s_{i,t})$, is multiplied by the return starting from the time step t rather than the return starting from time step 0 as done previously (see Figure 3.4).

3.2.1.3 Policy Gradient Theorem

It was proven by Sutton et al. [12] that we can estimate the policy gradient using the reward-to-go $\sum_{t'=t}^{T-1} \gamma^{t'-t} r(s_{t'}, a_{t'}, s_{t'+1})$ instead of the Q-values:

$$\nabla_\theta J(\theta) = \mathbb{E}_{s \sim \rho_\theta, a \sim \pi_\theta}[\nabla_\theta \log \pi_\theta(s,a) Q^{\pi_\theta}(s,a)], \tag{3.37}$$

where ρ_θ is the distribution of states encountered under policy π_θ. Reinforce algorithm can be seen as a special case of policy gradient theorem. In fact, the Q-value of an action is an estimate of the reward-to-go after performing that action.

Using the policy gradient theorem, the expectation is based on single transitions $\{(s_{i,t}, a_{i,t}, r_{i,t}, s_{i,t+1})\}$ rather than full trajectories and that enable the use of bootstrapping as in Temporal Difference methods (Section 2.6).

In practice, the true Q-values are unknown, so they should be estimated. It was proven in [12] that we can estimate the Q-values using a function approximator Q_φ with parameters φ and we still have an unbiased estimator of the policy gradient:

$$\nabla_\theta J(\theta) = \mathbb{E}_{s \sim \rho_\theta, a \sim \pi_\theta}[\nabla_\theta \log \pi_\theta(s,a) Q_\varphi(s,a)]. \tag{3.38}$$

The Q-value approximator should follow the *Compatible Function Approximation Theorem* which assert that the Q-values should be compatible with the

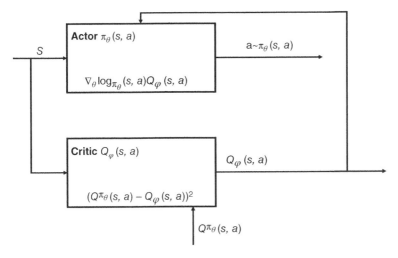

Figure 3.5 Actor-critic policy gradient architecture.

policy $\nabla_\varphi Q_\varphi(s, a) = \nabla_\theta \log \pi_\theta(s, a)$ and minimize the MSE with the true Q-values, $\mathbb{E}_{s \sim p_\theta, a \sim \pi_\theta}[(Q^{\pi_\theta}(s, a) - Q_\varphi(s, a))^2]$. However, in DRL algorithms, these conditions are not always met.

3.2.2 Actor-Critic Methods

The presentation of the Q-value approximators takes us to a new category of DRL algorithms, namely, the **actor-critic** algorithms, where the **actor** $\pi_\theta(s, a)$ learns the optimal policy and the **critic** learns to estimate the Q-values $Q_\varphi(s, a)$.

Most of the policy gradient DRL algorithms (e.g. A3C, DDPG, TRPO) adopt the actor-critic architecture as shown in Figure 3.5. Note that we can update the actor and critic networks using single transitions. Although REINFORCE algorithm is strictly an **on-policy** algorithm, the policy gradient methods can be transformed to **off-policy learning**, for instance, using the replay buffer described in DQN (Section 3.1.1) to store the sampled transitions. Additionally and similar to REINFORCE, the policy gradient theorem suffers from **high variance**. The next methods that we will introduce, will attempt to reduce this variance as well as the sample complexity using different techniques: advantages, deterministic policies, and natural gradients. One last remark is that the actor and the critic might be entirely separated or share some parameters (e.g. early layers such as the convolutional layers).

Example: The authors in [13] addressed the user scheduling problem in a multiuser massive multiple-input multiple-output (MIMO) wireless system.

Considering the channel conditions and the quality of service (QoS) demands, user scheduling handles the resource block allocation to base stations (BSs) and mobile users. A DRL-based coverage and capacity optimization method was proposed that dynamically returns the scheduling parameters indicators and a unified threshold of QoS metric. The state space is an indicator of the average spectrum efficiency, and the action space is a set of scheduling parameters. Given the average spectrum efficiency, the agent applies the selected scheduling parameters and receives a reward signal which is the average spectrum efficiency. The policy network is trained using a variant of the REINFORCE algorithm with trajectories generated from the current policy.

3.2.3 Advantage of Actor-Critic Methods

The policy gradient theorem introduces the actor-critic methods where the actor learns the optimal policy and the critic learns the true Q-values to estimate the policy gradient. In contrast to REINFORCE algorithm where the log-likelihood $\nabla_\theta \log \pi_\theta(a|s)$ is multiplied by the return computed using already sampled trajectory, the actor-critic methods can be updated using single-transitions. As in REINFORCE, the vanilla actor-critic method suffers from high variance; therefore, we need to use a baseline to reduce it. We choose a state-dependent baseline, which is precisely the state-value $V^\pi(s)$. The factor which is multiplied by the log-likelihood in the policy gradient is

$$A^\pi(s, a) = Q^\pi(s, a) - V^\pi(s), \tag{3.39}$$

where $A^\pi(s, a)$ is the advantage of using the action a under state s. In **Advantage Actor-Critic** methods, the advantage $A^\pi(s, a)$ is approximated by a parameterized function A_φ:

$$\nabla_\theta J(\theta) = \mathbb{E}_{s \sim \rho_\theta, a \sim \pi_\theta}[\nabla_\theta \log \pi_\theta(s, a) A_\varphi(s, a)], \tag{3.40}$$

where $A_\varphi(s, a)$ is equal to the expected advantage in expectation. There exist different ways to define the advantage estimate:

- **Monte Carlo advantage estimate**, where the Q-value $Q^{\pi_\theta}(s, a)$ is replaced by the actual return (i.e. reward-to-go) $A_\varphi(s, a) = R(s, a) - V_\varphi(s)$.
- **TD advantage estimate**, using an estimate of the actual return $A_\varphi(s, a) = r(s, a, s') + \gamma V_\varphi(s') - V_\varphi(s)$.
- **n-step advantage estimate**, $A_\varphi(s, a) = \sum_{k=0}^{n-1} \gamma^k r_{t+k+1} + \gamma^n V_\varphi(s_{t+n+1}) - V_\varphi(s_t)$.

Monte Carlo and temporal difference methods can be used as well, but the resulting algorithm will suffer from their disadvantages, i.e. the requirement of finite tasks, slow learning for Monte Carlo method and instability for TD method.

3.2.3.1 Advantage of Actor-Critic (A2C)

Advantage Actor-Critic (A2C) algorithm relies basically on the n-step method of computing the advantage. The n-step method can be seen as a mixture of TD and Monte Carlo methods. In fact, the Monte Carlo method estimates the Q-values from the reward-to-go (the sum of the rewards received after applying action a at state s under the current policy), and the TD method estimates the expected return by the state-value $V^\pi(s)$. The policy gradient expression is as follows:

$$\nabla_\theta J(\theta) = \mathbb{E}_{s \sim \rho_\theta, a \sim \pi_\theta} \left[\nabla_\theta \log \pi_\theta(s, a) \left(\sum_{k=0}^{n-1} \gamma^k r_{t+k+1} + \gamma^n V_\varphi(s_{t+n+1}) - V_\varphi(s_t) \right) \right].$$

(3.41)

The Q-value is estimated through n-step successive transitions $\sum_{k=0}^{n-1} \gamma^k r_{t+k+1}$ and the return afterward is estimated using the state-value of the $(n+1)$th state encountered $V^\pi(s_{t+n+1})$. This policy gradient estimation presents a trade-off between bias in TD with the wrong updates of the estimated values and variance in Monte Carlo, where there is high variability in the obtained returns. A2C is not totally different from the REINFORCE algorithm, and the basic steps remain the same: transition sampling, computing the n-step return, and then the policy updating. But it does not require finite tasks, and it has a critic that learns in parallel. A2C algorithm is summarized in **Algorithm 3.3**.

All the states are updated on the same number of transitions n. The terminal state in an episode will use only the final reward and the value of the terminal state; meanwhile, the first state in the episode uses the next n steps to compute the return. Although in practice, the discount factor γ does not really matter, it has a high influence on the results.

A2C operates in *online learning* because it explores a few transitions using the current policy and right after that the policy is updated. It does not require to wait till the end of an episode which makes it suitable for tasks with infinite horizon.

Similar to the value-based methods, the neural networks used for actor and critic will suffer from correlated input data that may appear in the minibatches since we are updating the neural networks on similar states. However, A2C does not require a replay buffer as in DQN. It relies on the use of multiple independent parallel actors to enhance the exploration and ensures independence between the states sampled. The concept is introduced in Asynchronous Advantage Actor-Critic (A3C), but applies also for A2C. Multiple threads (called **workers** or **actor-learners**) are created where each thread contains its own instance of the environment. At the start of each episode, each worker receives the most recent weights of the actor and critic from the global network. Then, the workers perform episodes independently starting from distinct states using different seeds

---

Algorithm 3.3 A2C algorithm for episodic task

1: **Input:** n-step size n, episode period T, learning step-size η

2: Initialize randomly actor network π_θ and the critic network V_φ.

3: Observe initial state s_0.

4: **for** $t = 0, \ldots, T$ **do**

5:     Initialize empty episode minibatch.

6:     **for** $k = 0, \ldots, n$ **do**

7:         Select action $a_k \sim \pi_\theta(s)$.

8:         Receive next state s_{k+1} and reward r_k.

9:         Store the transition (s_k, a_k, r_k, s_{k+1}) in the episode minibatch.

10:     **end for**

11:     if s_n is not terminal: set $R = V_\varphi(s_n)$ with critic, else $R = 0$.

12:     $d\theta \leftarrow 0, d\varphi \leftarrow 0$.

13:     **for** $k = n - 1, \ldots, 0$ **do**

14:         Update the discounted sum of rewards $R \leftarrow r_k + \gamma R$.

15:         Accumulate the policy gradient using the critic:

$$d\theta \leftarrow d\theta + \nabla_\theta \log \pi_\theta(s_k, a_k)(R - V_\varphi(s_k)). \tag{3.42}$$

16:         Accumulate the critic gradient:

$$d\varphi \leftarrow d\varphi + \nabla_\varphi (R - V_\varphi(s_k))^2. \tag{3.43}$$

17:     **end for**

18:     Update the actor and critic using SGD:

$$\theta \leftarrow \theta + \eta d\theta, \qquad \varphi \leftarrow \varphi - \eta d\varphi. \tag{3.44}$$

19: **end for**

---

so that they sample uncorrelated episodes. Subsequently, the workers accumulate the local gradients of the actor and critic networks and return them to the global network. Finally, the global network combines the received gradients from the workers and updates the global actor and critic networks so that they are sent back to the workers for the next episode, and so on, till convergence. The process is shown in **Algorithm 3.4**.

The use of parallel actors alleviates the problem of correlated input data by using different seeds or different exploration rates. However, this technique is not always available in real-world problems. For instance, it can be easily used in simulated environments such as video games, but in the robotics field, it is much harder where a brute-force solution involves multiple robots and lets them learn in parallel [15].

Algorithm 3.4 Parallel version of A2C [14]

1: Initialize the actor network π_θ and the critic V_φ in the global network.

2: **repeat**

3:     **for** each worker i in **parallel do**

4:         Get a copy of the global actor π_θ and critic V_φ.

5:         Sample an episode of n steps.

6:         Compute the accumulated gradients $d\theta_i$ and $d\varphi$.

7:         Send the accumulated gradients to the global network.

8:     **end for**

9:     Wait for all workers to terminate.

10:    Merge all accumulated gradients $d\theta$ and $d\varphi$.

11:    Update the global actor and critic networks.

12: **until** Satisfactory performance achieved

Example: The parallel version of A2C is suitable to solve problems involving distributed entities within the network. For instance, Heydari et al. [16] designed offloading policies for noncooperative mobile devices while optimizing the wireless resource allocation. The work adopted A2C to learn these policies since A2C can handle larger state and action spaces than DQN.

3.2.3.2 Asynchronous Advantage Actor-Critic (A3C)

A3C [17] enhances the A2C technique by eliminating the necessity for worker synchronization at the termination of each episode before applying the gradients. The argument is that each worker's task may need various time frames to accomplish; thus, they must be synchronized. Some workers may thus remain idle for the majority of the time, which results in a wastage of resources. Gradient merging and parameter updates are sequential processes; therefore, even if the number of workers is increased, no substantial *speedup* is predicted.

The key idea is simply to skip the synchronization: workers read and update the network parameters independently. For sure, the *concurrency* of data access for the reading and writing arises in A3C that may lead to two different learned policies. This kind of sharing of parameters between workers is called *HogWild!* updating [18], and it was shown to work effectively under certain conditions. We summarize the A3C method in **Algorithm 3.5**.

The workers operate fully independently, the only communication occurring between them is through the **asynchronous** updating of the global networks which lead to an efficient parallel implementation. For evaluation, Minh et al. [17] used A3C on Atari games using 16 CPU cores instead of graphics processing unit (GPU), and they achieved superior performance with reduced training time (1 day with 16 CPU cores and 8 days with GPU) than DQN that was trained

---

Algorithm 3.5 A3C pseudocode

1: Initialize the actor π_θ and the critic V_φ in the global network.
2: **for** each worker i **in parallel do**
3:     **repeat**:
4:         Get a local copy of the global networks π_θ and V_φ.
5:         Sample a trajectory of n transitions.
6:         Compute the accumulated gradients $d\theta_i$ and $d\varphi_i$.
7:         Update the global actor and critic network synchronously.
8:     **until** satisfactory performance achieved
9: **end for**

---

using GPU. The *speedup* is almost linear which means that with a high number of workers, we get better performance with fast computations. This is due to the fact that the policy is updated with more uncorrelated samples.

Entropy Regularization An additional feature in A3C is the entropy regularization added to the policy gradient to enforce the exploration during the learning. The entropy of a policy is expressed as follows [19]:

$$H(\pi_\theta(s_t)) = -\sum_a \pi_\theta(s_t, a_t) \log \pi_\theta(s_t, a). \tag{3.45}$$

For a continuous action space, the sum is replaced by an integral. Basically, the entropy measures the randomness of a policy. For fully deterministic policies where the same action is selected, the entropy is equal to zero because it brings no information about other actions. For fully uniform random policies, the entropy is maximal. Maximizing both the entropy and the returns would improve the exploration during learning by pushing the policy toward being nondeterministic. The policy gradient proposed in [20] is expressed as follows:

$$\nabla_\theta J(\theta) = \mathbb{E}_{s \sim \rho_\theta, a \sim \pi_\theta} \left[\nabla_\theta \log \pi_\theta(s, a)(R_t - V_\varphi(s)) + \beta \nabla_\theta H(\pi_\theta(s)) \right]. \tag{3.46}$$

The parameter β supervises the level of the regularization used. It is unfavorable to give too much of importance to the entropy that may lead to a uniformly random policy that does not solve the problem of maximizing the return. Also, we do not want to give too low importance that will take the policy to be suboptimal deterministic since there will be no exploration for better trajectories. Entropy regularization introduces a new hyperparameter β; nevertheless, it can be very advantageous if the hyperparameter is properly chosen.

Example: [21] proposed a DRL-based method to optimize the handover process that lowers the handover rate ensuring high-system throughput. Considering a network with multiple mobile users, small cell base stations (SBS), and one central controller, the agents represented by the mobile user select their serving SBS at each time slot minimizing the number of handover occurrences and ensuring a certain throughput. The states are defined by the reference signal quality received from the candidate SBSs and the last action selected. The actions are represented by the set of the SBSs selected. The reward function is defined by the difference between the data of the user and the energy consumption for the handover process. Due to the high density of the users, A3C was combined with long short-term memory (LSTM) to find the optimal policy within a short training time.

3.2.3.3 Generalized Advantage Estimate (GAE)

We recall the basic form of the policy gradient introduced at the beginning of the discussion of policy gradient techniques:

$$\nabla_\theta J(\theta) = \mathbb{E}_{s_t \sim \rho_\theta, a_t \sim \pi_\theta}[\nabla_\theta \log \pi_\theta(s_t, a_t)\psi_t], \tag{3.47}$$

where

- $\psi_t = R_t$, for the *REINFORCE* algorithm (Monte-Carlo sampling),
- $\psi_t = R_t - b$, for the *REINFORCE* with *baseline* algorithm,
- $\psi_t = Q^\pi(s_t, a_t)$, for the *policy gradient theorem*,
- $\psi_t = A^\pi(s_t, a_t)$, for the *advantage actor critic*,
- $\psi_t = r_{t+1} + \gamma V^\pi(s_{t+1}) - V^\pi(s_t)$, for the *TD actor critic*, and
- $\psi_t = \sum_{k=0}^{n-1} r_{t+k+1} + \gamma V^\pi(s_{t+n+1}) - V^\pi(s_t)$, for the *n-step* algorithm (A2C).

In general, the term ψ_t multiplied by the gradient of log-probabilities $\nabla_\theta \log \pi_\theta$ defines the expected return at the current state. The more ψ_t is based on real rewards, the more the gradient is correct in the average (low bias), the more it will vary (high variance). Therefore, we need a large number of samples to correctly estimate the gradient which increases the **sample complexity**. On the other hand, the higher the value of ψ_t is based on estimators as in TD error, the more stable the gradient is (low variance); however, the more faulty it is (high bias) which leads to an inferior policy. This issue is referred to as the *bias/variance* trade-off, which is a common trade-off in machine learning. The *n-step* method employed in A2C algorithm trades-offs between high variance and high bias. **Generalized Advantage Estimate** (GAE) [22] was introduced for an additional control over this trade-off between bias and variance. The main idea relies on merging the use of *advantages* and *n-step* method which results in *n-step advantage* defined as follows:

$$A_t^n = \sum_{k=0}^{n-1} \gamma^k r_{t+k+1} + \gamma^n V^\pi(s_{t+n+1}) - V^\pi(s_t)$$

$$= \sum_{l=0}^{n-1} \gamma^l \delta_{t+l}, \tag{3.48}$$

where $\delta_t = r_{t+1} + \gamma V^\pi(s_t)$ is the TD error. In GAE, rather than choosing *n-step* and tune the number of steps *n*, it simply averages all *n-step advantages* and weighs them with a discount parameter λ:

$$A_t^{\text{GAE}(\gamma,\lambda)} = (1 - \lambda) \sum_{l=0}^{\infty} \lambda^l A_t^l = \sum_{l=0}^{\infty} (\gamma\lambda)^l \delta_{t+l}. \tag{3.49}$$

The GAE can be seen as the discounted sum of all the *n*-step advantages. If $\lambda = 0$, we have the TD error method $A_t^{\text{GAE}(\gamma,0)} = \delta_t$ (high bias, low variance). And if $\lambda = 1$, we have the Monte-Carlo (MC) advantage $A_t^{\text{GAE}(\gamma,1)} = R_t$, i.e. the Monte Carlo advantage (low bias, high variance). Hence, varying λ between 0 and 1 controls the trade-off between bias and variance.

Note that γ and λ play different roles in the policy gradient. γ is the discount factor, which determines how much the future rewards are taken into consideration; hence, with high $\gamma < 1$, the bias is low, and variance is high. Schulman et al. [22] found empirically that taking λ smaller than γ (ordinarily equal to 0.99) leads to less bias. In conclusion, the policy gradient is expressed as follows:

$$\nabla_\theta J(\theta) = \mathbb{E}_{s_t \sim \rho^\pi, a_t \sim \pi_\theta} \left[\nabla_\theta \log \pi_\theta(s_t, a_t) \sum_{l=0}^{\infty} (\gamma\lambda)^l \delta_{t+l} \right]. \tag{3.50}$$

Schulman et al. [22] used *trust region optimization* as an additional enhancement in optimizing the policy gradient to reduce the bias which will discuss later in Section 3.4.2. We summarize the GAE method in **Algorithm 3.6**.

Algorithm 3.6 GAE pseudocode

1: **Input:** *n-step* size *n*, episode period *T*, learning step-size η.
2: Initialize the actor π_θ and the critic V_φ with random weights.
3: Observe initial state s_0.
4: **for** $t = 1, \ldots, T$ **do**
5:     Initialize empty minibatch.
6:     **for** $k = 0, \ldots, n$ **do**
7:         Select action $a_k \sim \pi_\theta(s)$. Receive the next state s_{k+1} and reward r_k.
8:         Store (s_k, a_k, r_k, s_{k+1}) in the minibatch.
9:     **end for**
10:     **for** $k = 0, \ldots, n$ **do**
11:         Compute the TD error $\delta_k = r_k + \gamma V_\varphi(s_{k+1}) - V_\varphi(s_k)$
12:     **end for**
13:     **for** $k = 0, \ldots, n$ **do**
14:         Compute the GAE advantage $A_k^{\gamma\lambda} = \sum_{l=0}^{\infty} (\gamma\lambda)^l \delta_{k+1}$
15:     **end for**
16:     Update the actor using the GAE advantage and natural gradients (TRPO).
17:     Update the critic using natural gradients (TRPO).
18: **end for**

Example: The expression of the advantages proposed by GAE is widely used in DRL-based methods in wireless communication. For instance, in [23], the authors proposed a distributed approach for contention-based spectrum access using a DRL-based method. The approach includes the estimation of the advantages which are performed through GAE.

3.3 Deterministic Policy Gradient (DPG)

The actors in the actor-critic methods described previously use stochastic policy $\pi_\theta(s)$: probabilities are assigned to the discrete actions and sampled from learned distribution for continuous actions. Stochastic policies are beneficial for the exploration process because in every state most of the actions have nonzero selection probability which let us to discover a wide range of state-action space. However, there are two major drawbacks. First, the policy gradient theorem provides **on-policy** methods: the value of an action $Q^\pi(s, a)$ estimated by the critic neural network $Q_\varphi(s, a)$ should be produced by the recent actor network $\mu_\theta(\cdot)$; otherwise, the bias would increase dramatically. Additionally, the **on-policy** algorithms prevent using the experience replay memory that stabilizes the learning as in DQN. Second, there is a relevant variance in the policy gradient because the optimal policy may generate two different returns in two episodes due to the stochasticity of the policy. Hence, the policy gradient methods are well known to have the worst **sample complexity** than the value-based methods, i.e. they require more samples of transitions to converge to the optimal policy and overcome the high variance issue.

On the other hand, the value-based methods provide a **deterministic policy** where the greedy policy is followed after learning. That is, the action selected is the action that has the highest state-action value $a^* = \arg\max_a Q_\theta(s, a)$. During the learning process, to ensure the exploration of new trajectories, the value-based methods adopt ϵ-greedy policy to introduce randomness in the selection of actions. Therefore, they consider **off-policy** methods because the learned policy is different from the one used during learning.

In this section, we introduce the distinguished **DDPG** algorithm that combines features of policy gradient (actor-critic architecture, continuous spaces, stability) and value-based methods (sample efficiency, off-policy).

3.3.1 Deterministic Policy Gradient Theorem

In [24], Silver et al. proposed a framework for policy gradient with deterministic policies rather than stochastic policies. We recall that the learning goal is to learn a policy that maximizes the expectation of the reward-to-go (expected return) over the achievable states under the current policy after every action:

$$\max_\theta J(\theta) = \mathbb{E}_{s \sim \rho_\mu}[R(s, \mu_\theta(s))], \tag{3.51}$$

where μ represents the parameterized deterministic policy and ρ_μ is the distribution of states accessible by the policy μ. Additional approximations should be performed because it is hard to estimate ρ_μ. In [24], Silver et al. proposed a usable policy gradient of deterministic policies.

Noting that the Q-values are the expectation of the reward-to-go after an action is taken $Q^\pi(s, a) = \mathbb{E}_\pi[R(s, a)]$, we can either maximize the true Q-values or the returns since both result in the same optimal policy. This idea is basic for dynamic programming where the *policy evaluation* searches the true Q-values of all state-action pairs, then the *policy improvement* updates the policy by selecting the action with the highest Q-value: $a_t^* = \arg\max_a Q_\theta(s_t, a_t)$.

For continuous spaces, the gradient of the objective function is the same as the gradient of the Q-function [24]. Hence, if we have an unbiased estimate $Q^\mu(s, a)$ of the Q-values, then updating the policy $\mu_\theta(s)$ in the direction of $\nabla_\theta Q^\mu(s, a)$ leads to an increase of the Q-values, and therefore, to an increase of the associated return.

$$\nabla_\theta J(\theta) = \mathbb{E}_{s \sim \rho_\mu} \left[\nabla_\theta Q^\mu(s, a)|_{a=\mu_\theta(s)} \right]. \tag{3.52}$$

Using the chain rules, we expand the expression of the gradient of the Q-function:

$$\nabla_\theta J(\theta) = \mathbb{E}_{s \sim \rho_\mu} \left[\nabla_\theta \mu_\theta(s) \times \nabla_a Q^\mu(s, a)|_{a=\mu_\theta(s)} \right]. \tag{3.53}$$

From Eq. (3.53), which shows the main characteristic of the **actor-critic** architecture, we observe that the first term $\nabla_\theta \mu_\theta(s)$ represents the gradient of the parameterized actor and the second term $\nabla_a Q^\mu(s, a)$ represents the critic, which tells the actor in which direction to update the policy, the direction of actions that are associated with higher rewards.

Similar to stochastic policy gradient, the challenge at this point is obtaining the unbiased estimate of $Q^\mu(s, a)$ and computing its gradient. In [24], the authors showed that a parameterized function approximator $Q_\varphi(s, a)$ can be useful if it is compatible and minimizing the quadratic error with the true Q-values:

$$\nabla_\theta J(\theta) = \mathbb{E}_{s \sim \rho_\mu} \left[\nabla_\theta \mu_\theta(s) \times \nabla_a Q_\varphi(s, a)|_{a=\mu_\theta(s)} \right], \tag{3.54}$$

$$\mathcal{L}(\varphi) = \mathbb{E}_{s \sim \rho_\mu} \left[\left(Q^\mu(s, \mu_\theta(s)) - Q_\varphi(s, \mu_\theta(s)) \right)^2 \right]. \tag{3.55}$$

Silver et al. [24] proved that **Deterministic Policy Gradient** (DPG) with linear function approximators performs similar to the stochastic algorithms in high-dimensional or continuous action spaces.

3.3.2 Deep Deterministic Policy Gradient (DDPG)

To extend the DPG with nonlinear function approximators, Lillicrap et al. [25] proposed a very successful algorithm, namely, **DDPG**, to solve problems with continuous action space. Basically, DDPG combines two key concepts: DQN and DPG. From DQN, it uses the concepts of **replay buffer** to learn in off-policy by storing the old transitions, and the **target networks** that stabilize the learning process. Moreover, in the DQN, the target networks are updated every couple of thousands of steps. It is more efficient to let the target networks to follow the actual networks using a sliding average for the actor and critic networks in each training step:

$$\theta' \leftarrow \tau\theta + (1 - \tau)\theta' \quad \text{(polyak averaging)}, \tag{3.56}$$

where $\tau \ll 1$. The target networks are a late version of the trained network, and hence, they stabilize the learning of the Q-values.

On the other hand, DDPG exploits the expression of the policy gradient from DPG. The critic network is trained similar to Q-learning with the help of the critic target network:

$$\mathcal{L}(\varphi) = \mathbb{E}_{s \sim \rho_\mu} \left[\left(r(s, a, s') + \gamma Q_{\varphi'}(s', \mu_{\theta'}(s)) - Q_\varphi(s, a) \right)^2 \right]. \tag{3.57}$$

Since the policy is deterministic, it can always return the same actions and may miss other actions that might lead to higher returns. For stochastic policies, all the actions are assigned with a nonzero selection probability, so the exploration is implicit in the policy. The solution proposed with DDPG algorithm is to add some noise to the action produced by the policy:

$$a_t = \mu_\theta(s_t) + \xi. \tag{3.58}$$

The **additive noise** can be any random process (e.g. Gaussian noise). The original paper on DDPG [25] suggested to use **Ornstein–Uhlenbeck** process [26]: it generates temporally correlated noise with zero mean. They are employed in physics for modeling the velocity of Brownian particles with friction. The pseudocode for DDPG is summarized in **Algorithm 3.7**.

DDPG is an off-policy algorithm because the actual actor training samples μ_θ are produced by an outdated version of the target policy $\mu_{\theta'}$. It has become one of the most popular model-free DRL algorithms to solve problems with continuous action spaces. As stated in the original paper, *batch normalization* [27] can significantly stabilize the training of neural networks. The main limitation of DDPG is the high sample complexity required for the algorithm to converge to a near-optimal policy.

Algorithm 3.7 DDPG algorithm

1: Initialize actor network μ_θ and critic network Q_φ with random weights.
2: Initialize actor target network $\mu_{\theta'}$ and critic target networks $Q_{\varphi'}$ with same weights of the actual networks $\theta'' \leftarrow \theta, \varphi' \leftarrow \varphi$.
3: Initialize experience replay memory (ERM) D.
4: **for** each episode **do**
5:     Initialize random process ξ.
6:     Observe initial state s_0.
7:     **for** $t = 1, \dots, T$ **do**
8:         Select action $a_t = \mu_\theta(s_t) + \xi_t$.
9:         Perform action a_t and observe the new state s_t' and reward r_t.
10:        Store the transition (s_t, a_t, r_t, s_t') in the ERM B.
11:        Sample minibatch of N transitions from B.
12:        **for** each transition (s_k, a_k, r_k, s_k') in the minibatch **do**
13:           Compute the target value using the target networks:

$$y_k = r_k + \gamma Q_{\varphi'}(s_k', \mu_{\theta'}(s_k')). \tag{3.59}$$

14:        **end for**
15:        Update the critic network by minimizing:

$$\mathcal{L}(\varphi) = \frac{1}{N} \sum_k \left(y_k - Q_\varphi(s_k, a_k) \right)^2. \tag{3.60}$$

16:        Update the actor network using the sampled policy gradient:

$$\nabla_\theta J(\theta) = \frac{1}{N} \sum_k \nabla_\theta \mu_\theta(s_k) \times \nabla_a Q_\varphi(s_k, a)|_{a=\mu_\theta(s_k)}. \tag{3.61}$$

17:        Update the target networks:

$$\theta' \leftarrow \tau\theta + (1 - \tau)\theta', \tag{3.62}$$

$$\varphi' \leftarrow \tau\varphi + (1 - \tau)\varphi'. \tag{3.63}$$

18:    **end for**
19: **end for**

Example: DDPG has been widely used in wireless communication problems specifically for problems, where the action space is continuous. For example, in [28], the authors used DDPG to solve the beamforming optimization problem in a cell-free MIMO network. Given the states defined as the signal-to-interference-plus-noise ratios (SINRs) of the users, the agent predicts the beamforming elements of the users and receives a reward signal, which is the normalized-sum rated chosen as a performance metric of the network.

3.3.3 Distributed Distributional DDPG (D4PG)

Basically, it is a **distributed** version of DDPG that includes numerous improvements to enhance its efficiency [29] as follows:

- The critic training that is based on **distributional learning** [30] where the critic learns the distribution of Q-values rather learns an estimator of the Q-values which reduces variance.
- **n-step** method that is used in the computation of the target Q-values.
- **Multiple distributed parallel actors** that collect independent transitions in parallel and store them in a global replay buffer.
- Using a PER (Section 3.1.3) [7] in place of the classical replay buffer and sampling the transitions based on the information gain rather than uniformly.

3.4 Natural Gradients

So far, the neural networks have been used in DRL as function approximators, and they are optimized using **SGD** or its variants such as RMSProp and Adam. SGD minimizes the loss function by updating the parameters θ in the opposite direction of the gradient of the loss function (gradient descent), or in the same direction for policy gradient (gradient ascent), proportionally to learning rate η:

$$\theta \leftarrow \theta - \eta \nabla_\theta \mathcal{L}(\theta) \qquad \text{(gradient descent)}. \tag{3.64}$$

SGD is also referred to as a **steepest descent method**, where it looks for the lowest change of parameters that give the largest negative change in the loss function. This is desirable in a supervised learning setting where we seek to minimize a loss function in the fastest possible way while holding the changes in the weights small enough to stabilize the learning. This is because the changes in weights due to a single minibatch may erase the weights obtained due to the previous minibatches. The other difficulty is the tuning of the learning rate: a too high rate would lead to an unstable learning, while a too low rate would make the convergence of the learning process very slow.

In DRL methods, we have an additional difficulty which is the nonstationarity of the target which is used to train the critic/Q-function. For instance, in Q-learning, the target value $r(s, a, s') + \gamma \max_{a'} Q_\theta(s', a')$ changes along with the parameters θ. If the Q-values vary widely from a minibatch to another one, the neural network will have an unstable target to converge to, and thus, we will get a suboptimal policy. This is where the **target network** comes to play: they are used to compute the target Q-values. They can be an old version of the trained network updated every couple of thousands of steps as in DQN, or a smoothed version as in DDPG. It would introduce biases in the Q-values because the targets will be continuously faulty; however, this bias vanishes after sufficient update steps, but at the cost of high sample complexity.

For **on-policy** methods, we cannot use the target networks. The critic network should be always updated using transitions that are generated by the most recent version of the actor. This may lead to a huge waste of data because the old transitions will not be used in the update of the actor and critic networks. The policy gradient theorem shows the reason behind that

$$
\begin{aligned}
\nabla_\theta J(\theta) &= \mathbb{E}_{s\sim\rho_\theta, a\sim\pi_\theta}[\nabla_\theta \log \pi_\theta(s,a) Q^{\pi_\theta}(s,a)] \\
&\approx \mathbb{E}_{s\sim\rho_\theta, a\sim\pi_\theta}[\nabla_\theta \log \pi_\theta(s,a) Q_\varphi(s,a)].
\end{aligned}
\tag{3.65}
$$

If the policy π_θ makes a big change, the estimated Q-function $Q_\varphi(s,a)$ corresponds to the Q-values of an entirely different policy rather than the true Q-values $Q^{\pi_\theta}(s,a)$. The estimated Q-values will be biased, and the learned policy will be suboptimal. Precisely, the actor should not change much faster than the critic. The naive solution for this is the use of a slow learning rate for the critic network; however, obviously, this will make the learning process taking forever in addition to the increase of the sample complexity.

To solve this problem, we need to update the parameters in the opposite way of the steepest descent. We want to look for the highest change of parameters that give us the lowest change in the policy, but still in the right direction. With a large change of the parameters, we will learn much more from each experience. With a small change in the policy, the past experience will be reusable experiences and the target values will not be too far from true values.

Natural gradients come to solve this problem, which is a statistical tool to optimize over probability distribution spaces. Historically, [31] included the natural gradients in the training of neural networks. Later on, in [32], the authors applied natural gradients to policy gradient methods. In [33], a natural actor-critic algorithm was proposed for linear function approximators. Natural Gradients is the core idea adopted in **Trust Region Policy Optimization** (TRPO) algorithm [34] and **Proximal Policy Optimization** (PPO) [35] which has recently taken the lead over DDPG to solve RL problems with continuous action space thanks to the reduced sample complexity and its robustness to the hyperparameters.

3.4.1 Principle of Natural Gradients

We consider two pairs of Gaussian distributions: the first pair is $\mathcal{N}(0, 0.2)$ and $\mathcal{N}(1, 0.2)$ (Figure 3.6) and the second pair is $\mathcal{N}(0, 10)$ and $\mathcal{N}(1, 10)$ (Figure 3.7). The distance in the Euclidean space of parameters $d = \sqrt{(\mu_1 - \mu_2)^2 + (\sigma_1 - \sigma_2)^2}$ of the first pairs is the same as of the second pair. However, obviously, the distributions in the first pair are far away from each other, and in the second pair, they are more closer to each other. With one sample, for the second pair of Gaussian

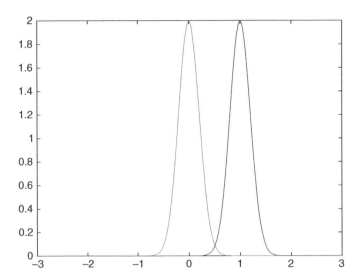

Figure 3.6 Gaussian distributions $\mathcal{N}(0, 0.2)$ and $\mathcal{N}(1, 0.2)$.

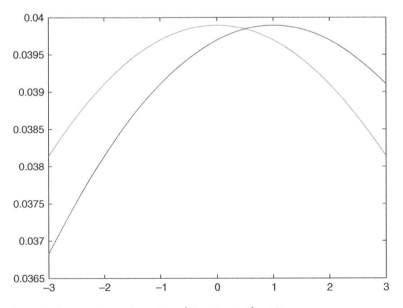

Figure 3.7 Gaussian distribution $\mathcal{N}(0, 10)$ and $\mathcal{N}(1, 10)$.

distributions, it is hard to distinguish which distribution it belongs to. Therefore, we cannot consider the Euclidean distance a rightful measurement of the distance between distributions.

In statistics, **Kullback–Leibler (KL) divergence** $D_{KL}(p||q)$ divergence is frequently used as a measure of the statistical distance between two distributions p and q. It is also called the *relative entropy* or the *information gain* and defined as follows:

$$D_{KL}(p||q) = \mathbb{E}_{x \sim p} \left[\log \frac{p(x)}{q(x)} \right] = \int p(x) \log \frac{p(x)}{q(x)} dx. \tag{3.66}$$

The minimum of KL divergence is 0 when $p = q$ since $\log \frac{p(x)}{q(x)}$ equal to 0, and positive otherwise. Minimizing D_{KL} is the same as matching two distributions. Moreover, all the supervised learning methods might be seen as a process of minimizing the KL divergence: let $p(x)$ be the distribution of the labels of the data x and $q(x)$ the distribution of the output of the model, then the learning leads to matching the distribution of the target to the distribution of the output of the model. This is the core concept used in *generative adversarial networks* [36] or *variational autoencoders* [37].

However, KL divergence is not symmetric $(D_{KL}(p||q) \neq D_{KL}(q||p))$. A more useful measurement of the divergence is a symmetric version of KL divergence, also called *Jensen–Shannon (JS)* divergence:

$$D_{JS}(p||q) = \frac{D_{KL}(p||q) + D_{KL}(q||p)}{2}. \tag{3.67}$$

There exist other divergence measurements, for instance *Wasserstein distance* that is used in generative adversarial networks (GANs) [36], but they are not relevant here.

Considering a parameterized distribution $p(x; \theta)$, the new parameterized distribution after applying parameter change $\Delta\theta$ is $p(x; \theta + \Delta)$. The Euclidean metric in the parameter space does not take the structure of the curvatures in the distributions; therefore, we define a *Riemannian* metric that takes into account the local curvature of the distribution between θ and $\Delta\theta$. In the case of using the symmetric KL divergence D_{JS}, the Riemann metric complies with the **Fisher Information Matrix**, which is defined by the hessian matrix of the KL divergence around θ:

$$F(\theta) = \nabla_\theta^2 D_{JS}(p(x; \theta)||p(x; \theta + \Delta\theta)). \tag{3.68}$$

The Fisher matrix includes the computation of the second-order derivatives that are slow to obtain, especially with a large number of parameters. A simple form of the Fisher matrix is defined by the expectation of the outer product between the gradient of the log-likelihoods:

$$F(\theta) = \mathbb{E}_{x \sim p(x; \theta)} \left[\nabla_\theta \log p(x; \theta)(\nabla_\theta \log p(x; \theta))^T \right]. \tag{3.69}$$

The Fisher matrix allows to approximate locally the KL divergence between two distributions:

$$D_{JS}\left(p(x;\theta)\|p(x;\theta+\Delta\theta)\right) \approx \Delta\theta^T F(\theta)\Delta\theta. \tag{3.70}$$

We can see that the symmetric KL divergence is quadratic in local regions ($\Delta\theta$ too small), and this means that the resulting update rules will be linear when we minimize the KL divergence using gradient descent.

The main idea of **natural gradient descent** is moving along the statistical manifold defined by the probability distribution p minimizing the loss function denoted by $\mathcal{L}$, by using the local curvature of the KL divergence surface, i.e. taking the direction $\overline{\nabla}_\theta \mathcal{L}(\theta)$, where

$$\overline{\nabla}_\theta \mathcal{L}(\theta) = F(\theta)^{-1}\nabla_\theta \mathcal{L}(\theta). \tag{3.71}$$

$\overline{\nabla}_\theta \mathcal{L}(\theta)$ is called the **natural gradient** of $\mathcal{L}(\theta)$. The natural gradient descent moves simply in the direction of the natural gradient:

$$\theta^{k+1} = \theta^k - \eta \overline{\nabla}_\theta \mathcal{L}(\theta). \tag{3.72}$$

In case the distribution is not curved, the natural gradient descent becomes equivalent to the regular gradient descent. Since the natural gradient relies on the inverse of the fisher matrix (curvatures), the magnitude of the natural gradient is high for flat regions that lead to larger steps and smaller steps for steep regions.

Natural gradient descent is an optimization method, which can efficiently train deep networks in supervised learning problems [38]. However, it requires the computation of the inverse of the Fisher information matrix with size depending on the number of free parameters.

3.4.2 Trust Region Policy Optimization (TRPO)

TRPO was proposed by Schulman et al. [34] by applying the natural gradients on the policy gradient theorem using nonlinear function approximators such as neural networks. To explain the ideas behind the TRPO, we recall the main goal of learning in DRL to maximize the expected return which we express as follows:

$$\eta(\pi) = \mathbb{E}_{s\sim\rho_\pi, a\sim\pi}\left[\sum_{t=0}^{\infty}\gamma^t r(s_t, a_t, s_{t+1})\right], \tag{3.73}$$

where ρ_π defines the probability distribution of states to be visited at some point in time under the policy π:

$$\rho_\pi(s) = P(s_0 = s) + \gamma P(s_1 = s) + \gamma^2 P(s_2 = s) + \dots. \tag{3.74}$$

Moreover, we can relate the expected return to two different policies π_θ and $\pi_{\theta_{old}}$ using advantages [39]:

$$\eta(\theta) = \eta(\theta_{old}) + \mathbb{E}_{s\sim\rho_{\pi_\theta}, a\sim\pi_\theta}[A_{\pi_{\theta_{old}}}(s, a)], \tag{3.75}$$

where $A_{\pi_{\theta_{old}}}(s,a)$ expresses the variation of the expected return in comparison with the expected return of an older policy. This expression is interesting because it measures on average how good a policy is compared to another policy. In order to compute this expression of expected return $\eta(\theta)$, we need to estimate the mathematical expectation that depends on the state-action pairs under the new policy ($a \sim \pi_\theta$ and $s \sim \rho_{\pi_\theta}$), and this is difficult to do. Alternatively, we approximate the expression for the expected return in the region where the policies π_θ and $\pi_{\theta_{old}}$ are close to each other. Hence, we can sample the states from the old distribution:

$$\eta(\theta) = \eta(\theta_{old}) + \mathbb{E}_{s \sim \rho_{\pi_{\theta_{old}}}, a \sim \pi_{\theta_{old}}}[A_{\pi_{\theta_{old}}}(s,a)]. \tag{3.76}$$

The main role of natural gradients is to update the parameters of the policy in the right direction while keeping the changes in the distribution as small as possible in order to make the assumption of the approximation valid. The objective function is defined as follows:

$$J_{\theta_{old}}(\theta) = \eta(\theta_{old}) + \mathbb{E}_{s \sim \rho_{\pi_{\theta_{old}}}, a \sim \pi_\theta}[A_{\pi_{\theta_{old}}}(s,a)]. \tag{3.77}$$

When $\theta = \theta_{old}$, we get $J_{\theta_{old}} = \eta(\theta_{old})$, and therefore, the gradient of $J_{\theta_{old}}$ w.r.t θ taken at θ_{old} is equal to $\nabla_\theta \eta(\theta_{old})$:

$$\nabla_\theta J_{\theta_{old}}|_{\theta=\theta_{old}} = \nabla_\theta \eta(\theta)|_{\theta=\theta_{old}}. \tag{3.78}$$

Hence, locally maximizing $J_{\theta_{old}}(\theta)$ w.r.t. θ is the same as maximizing the expected return $\eta(\theta)$. $J_{\theta_{old}}$ is called the **surrogate objective function**: it is not explicitly the objective function that we want to maximize, but its maximization leads to the same result.

Let us consider that we have the policy π' that maximizes $J_{\theta_{old}}$, i.e. that maximizes the advantage $A_{\pi_{\theta_{old}}}$ over old policy π_{old} for every state-action pair. We cannot guarantee that π' and $\pi_{\theta_{old}}$ are close to each other so that the assumption remains valid. In [39], Kakade and Langford proposed the *conservative policy iteration* method where the policy is updated by taking small steps in the direction of π':

$$\pi_\theta(s,a) = (1-\alpha)\pi_{\theta_{old}}(s,a) + \alpha\pi'(s,a). \tag{3.79}$$

Alternatively, we can penalize the objective function using the KL divergence to maintain the assumption valid. There are two ways of penalization:

i. By adding a hard constraint to the problem using the KL divergence between the policy distributions. The constrained optimization can be solved using Lagrange methods:

$$\begin{aligned} &\max_\theta J_{\theta_{old}}(\theta), \\ &\text{subject to: } D_{KL}(\pi_{\theta_{old}} || \pi_\theta) \leq \delta, \end{aligned} \tag{3.80}$$

where δ is a threshold.

ii. By adding a penalization term to the objective function (soft constraint):

$$\max_{\theta} \mathcal{L}(\theta) = J_{\theta_{old}}(\pi_{\theta}) - C\, D_{KL}(\pi_{\theta_{old}} || \pi_{\theta}), \tag{3.81}$$

where C is a constant and considered as a hyperparameter.

Both of the penalized problem formulations aim to find the policy that maximizes the expected return while ensuring it is still close enough to the current policy in terms of KL divergence. Additionally, the methods are highly sensitive to the choice of δ and C. The first method forces the KL divergence to be below a threshold δ, while the second method attempts to maximize $J_{\theta_{old}}$ and minimize the KL divergence at the same time.

Mathematically, the KL divergence $D_{KL}(\pi_{\theta_{old}} || \pi_{\theta})$ has to be the maximum over the state space:

$$D_{KL}^{\max}(\pi_{\theta_{old}} || \pi_{\theta}) = \max_{s} D_{KL}(\pi_{\theta_{old}}(s, .) || \pi_{\theta}(s, .)). \tag{3.82}$$

However, computing the maximum of KL divergence over the state space is infeasible. Therefore, we replace it with the mean KL divergence [34]:

$$\begin{aligned} \overline{D}_{KL}(\pi_{\theta_{old}} || \pi_{\theta}) &= E_s[D_{KL}(\pi_{\theta_{old}}(s, \cdot) || \pi_{\theta}(s, \cdot))] \\ &\approx \frac{1}{N} \sum_{i=1}^{N} D_{KL}(\pi_{\theta_{old}}(s_i, \cdot) || \pi_{\theta}(s_i, \cdot)). \end{aligned} \tag{3.83}$$

3.4.2.1 Trust Region

Since it not possible to compute or estimate the real objective function $\eta(\theta)$, we use a surrogate objective function $\mathcal{L}(\theta) = J_{\theta_{old}}(\theta) - C D_{KL}(\pi_{\theta_{old}} || \pi_{\theta})$. We already have that the two objective functions are equal in θ_{old}:

$$\mathcal{L}(\theta_{old}) = J_{\theta_{old}}(\theta_{old}) - C D_{KL}(\pi_{\theta_{old}} || \pi_{\theta_{old}}) = \eta(\theta_{old}). \tag{3.84}$$

Therefore, they have the same gradients in θ_{old}:

$$\nabla_{\theta} L(\theta)|_{\theta=\theta_{old}} = \nabla_{\theta} \eta(\theta)|_{\theta=\theta_{old}}. \tag{3.85}$$

Also, the surrogate objective function represents a lower bound of the real objective function:

$$\eta(\theta) \geq J_{\theta_{old}}(\theta) - C D_{KL}(\pi_{\theta_{old}} || \pi_{\theta}). \tag{3.86}$$

The most interesting part of natural gradients is that the values of θ which maximize $\mathcal{L}(\theta)$ represent a big step in the parameter space in the maximization of $\eta(\theta)$ since θ and θ_{old} are too different. And at the same time, it represents a small step in the distribution space of the policies $\pi_{\theta_{old}}$ and π_{θ} thanks to the constraint applied using the KL divergence. Thus, the parameter region around θ_{old} where the KL divergence is too small is called **trust region**. In other words, it is the region where we can take big optimization steps without risking of violation of the earlier assumption of the approximation.

3.4.2.2 Sample-Based Formulation

Although Schulamn et al. [34] used the regularized optimization in their theoretical proofs, the practical implementation uses the constrained optimization problem:

$$\max_{\theta} J_{\theta_{\text{old}}}(\theta) = \eta(\theta_{\text{old}}) + \mathbb{E}_{s \sim \rho_{\pi_{\text{old}}}, a \sim \pi_{\theta}}[A_{\pi_{\theta_{\text{old}}}}(s, a)],$$

$$\text{subject to: } D_{\text{KL}}(\pi_{\theta_{\text{old}}} \| \pi_{\theta}) \leq \delta. \tag{3.87}$$

Here, $\eta(\theta_{\text{old}})$ is considered as a constant in the maximization of $J_{\theta_{\text{old}}}(\theta)$ because it does not depend on θ. We are only interested in maximizing the advantage of actions taken by π_{θ} in every state visited by $\pi_{\theta_{\text{old}}}$. However, π_{θ} is the policy we are looking for, so we cannot sample from it. Therefore, we use the **importance sampling** to be able to use the old policy $\pi_{\theta_{\text{old}}}$ for sampling, which is possible as long as we modify the change the objective function accordingly with the *importance sampling* weight:

$$\max_{\theta} \mathbb{E}_{s \sim \rho_{\pi_{\text{old}}}, a \sim \pi_{\theta_{\text{old}}}} \left[\frac{\pi_{\theta}(s, a)}{\pi_{\theta_{\text{old}}}(s, a)} A_{\pi_{\theta_{\text{old}}}}(s, a) \right],$$

$$\text{subject to: } D_{\text{KL}}(\pi_{\theta_{\text{old}}} \| \pi_{\theta}) \leq \delta. \tag{3.88}$$

Since we are allowed to use sampling from the old policy now, we can safely generate many transitions using $\pi_{\theta_{\text{old}}}$, compute the advantages of all state-action pairs using the real rewards, then optimize the surrogate objective function using second-order optimization methods.

Alternatively, we can replace the advantages by the Q-values of the state-action pairs. In fact, the advantages $A_{\pi_{\theta_{\text{old}}}}(s, a) = Q_{\pi_{\theta_{\text{old}}}}(s, a) - V_{\pi_{\theta_{\text{old}}}}(s)$ depend on the state values which do not depend on the policies. Therefore, they are considered to be constants during each optimization step, and they can be removed safely:

$$\max_{\theta} \mathbb{E}_{s \sim \rho_{\pi_{\text{old}}}, a \sim \pi_{\theta_{\text{old}}}} \left[\frac{\pi_{\theta}(s, a)}{\pi_{\theta_{\text{old}}}(s, a)} Q_{\pi_{\theta_{\text{old}}}}(s, a) \right],$$

$$\text{subject to: } D_{\text{KL}}(\pi_{\theta_{\text{old}}} \| \pi_{\theta}) \leq \delta. \tag{3.89}$$

TRPO solves this optimization problem in every step.

3.4.2.3 Practical Implementation

To solve the constrained optimization problem, we can use the *Lagrange* method with the variable λ:

$$\mathcal{L}(\theta, \lambda) = J_{\theta_{\text{old}}}(\theta) - \lambda(D_{\text{KL}}(\pi_{\theta_{\text{old}}} \| \pi_{\theta}) - \delta). \tag{3.90}$$

The Lagrangian form is close to the regularized form introduced in Eq. (3.81). Later on, we use the second-order approximation of the KL divergence using the

Fisher Information Matrix:

$$D_{\text{KL}}(\pi_{\theta_{\text{old}}} \| \pi_\theta) = (\theta - \theta_{\text{old}})^T F(\theta_{\text{old}})(\theta - \theta_{\text{old}}). \tag{3.91}$$

Therefore, the Lagrangian function is expressed as follows:

$$\mathcal{L}(\theta, \lambda) = \nabla_\theta J_{\theta_{\text{old}}}(\theta - \theta_{\text{old}}) - \lambda(\theta - \theta_{\text{old}})^T F(\theta_{\text{old}})(\theta - \theta_{\text{old}}). \tag{3.92}$$

We can notice that it is quadratic in $\Delta\theta = \theta - \theta_{\text{old}}$; therefore, it has a unique maximum characterized by a zero first-order derivative:

$$\nabla_\theta J_{\theta_{\text{old}}}(\theta) = \lambda F(\theta_{\text{old}})\Delta\theta. \tag{3.93}$$

This leads us to the expression of the $\Delta\theta$:

$$\Delta\theta = \frac{1}{\lambda} F(\theta_{\text{old}})^{-1} \nabla_\theta J_{\theta_{\text{old}}}(\theta). \tag{3.94}$$

The expression of $\Delta\theta$ summarizes the natural gradient descent. It still depends on the size of the step $\frac{1}{\lambda}$ that should be determined, but it can be replaced by a fixed hyperparameter.

The computation of the inverse of the Fisher Information Matrix is expensive, which is quadratic with the number of parameters θ. In the original paper [34], the authors proposed the use of a *conjugate gradient* algorithm followed by a *line search* to find the new parameters θ_{k+1} that ensure the KL divergence constraint is satisfied.

In summary, TRPO is characterized by the following:

- policy gradient method.
- monotonically improving the expected return.
- using a surrogate objective function that represents the lower bound to iteratively update the parameters of the policy without changing the policy a lot (thanks to the KL divergence).
- less sensitive to the learning rate compared to DDPG.

On the other hand, it is difficult to use TRPO with neural networks with multiple outputs such as policy and value function because it is based on the policy distribution and its association with the parameters. Also, it performs well for fully connected layers; meanwhile, it performs poorly with convolutional neural networks (CNNs) and recurrent neural networks (RNNs). Finally, the implementation of the conjugate gradient is not straightforward as for the SGD.

Example: TRPO is considered as one of the policy-based algorithms that can be applied to discrete and continuous action spaces [40]. Valadarsky et al. [41] applied TRPO algorithm to learn optimal routing strategies in a network. The agent is represented by the system/operator. It chooses the routing strategy that should be adopted in the network. The state space is defined by the traffic demands between sources and destinations, and the reward function is the ration between the achieved performance and the optimal performance.

3.4.3 Proximal Policy Optimization (PPO)

To overcome the weakness of TRPO such as its nonefficiency while dealing with complex neural network architectures (e.g. CNN, RNN) and to increase the range of environments and tasks that it can solve, Schulman et al. proposed the **PPO**. The main idea is to change the surrogate objective function, i.e. the lower bound to expected return, to a simpler expression that can be solved using the first-order optimization techniques (e.g. gradient ascent). First, we rewrite the surrogate objective function of TRPO indicating the time index and the sampling weight ρ_t:

$$\mathcal{L}^{\text{CPI}}(\theta) = \mathbb{E}_t \left[\frac{\pi_\theta(s_t, a_t)}{\pi_{\theta_{\text{old}}}(s_t, a_t)} A_{\pi_{\theta_{\text{old}}}}(s_t, a_t) \right] = \mathbb{E}_t \left[\rho_t(\theta) A_{\pi_{\theta_{\text{old}}}}(s_t, a_t) \right]. \tag{3.95}$$

The term "CPI" refers to *conservative policy iteration* [39]. In the absence of the KL constraint, the policy updates will be too large which leads to huge instability. The authors aimed to modify the objective function and penalize the policy changes that make the sampling weight ρ_t too different from 1. In other words, they penalize the policy changes where the KL divergence is high. The objective function proposed is expressed as follows:

$$\mathcal{L}^{\text{CLIP}}(\theta) = \mathbb{E}_t \left[\min \left(A_{\pi_{\theta_{\text{old}}}}(s_t, a_t), \text{clip}(\rho_t(\theta), 1 - \epsilon, 1 + \epsilon) A_{\pi_{\theta_{\text{old}}}}(s_t, a_t) \right) \right]. \tag{3.96}$$

In the min function, the left part is the same as the surrogate function proposed in TRPO. The right part bounds the importance sampling weight to be around 1.

As shown in Figure 3.8, when the advantage $A(s_t, a_t) > 0$, i.e. selecting action a_t leads to a higher return than the expected action, the probability of selection

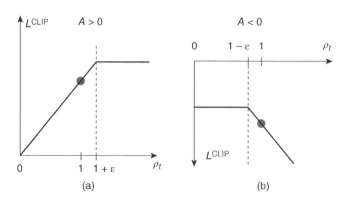

Figure 3.8 Clipping effect. (a): $A > 0$. (b): $A < 0$. Schulman et al. [35]/ arXiv / CC BY 4.0.

Algorithm 3.8 PPO pseudoalgorithm

1: Initialize randomly the actor π_θ and the critic V_φ.

2: **while** not converged **do**

3:     **for** N actors in parallel **do**

4:         Collect T transitions using old policy $\pi_{\theta_{old}}$.

5:         Compute the generalized advantage of each transition using the critic.

6:     **end for**

7:     **for** For K epochs **do**

8:         Sample M transitions from the previously collected.

9:         Train the actor to maximize the clipped surrogate objective.

10:        Train the critic to minimize the MSE using TD learning.

11:     **end for**

12:     $\theta_{old} \leftarrow \theta$.

13: **end while**

---

will increase $\pi_\theta(s_t, a_t) > \pi_{\theta_{old}}(s_t, a_t)$. Also, the sampling weight may increase highly (a change of probability from 0.01 to 0.05 has a sampling weight equal to $\rho_t = 5$). In such case, we clip the sampling weight to $1 + \epsilon$, so that the parameters will be updated in the right direction without changing a lot the policy. On the other hand, when the advantage is negative, the selection probability will be decreased and the sampling weight may become much smaller than 1, so we clip it by $1 - \epsilon$ to avoid the drastic changes of the policy. Then, we apply the min function to make sure the final objective function to be a lower bound of the unclipped function. Originally, the **Generalized Advantage Estimation** was proposed to estimate $A_{\pi_{\theta_{old}}}(s_t, a_t)$, but we still can use other methods such as n-steps method. Moreover, the sampled transitions are collected from multiple actors in parallel as in A2C and A3C to ensure that they are uncorrelated. **Algorithm 3.8** shows the pseudocode for **PPO**.

PPO is simpler than the TRPO: the clipped surrogate objective function can be directly maximized using first-order optimization techniques such as stochastic gradient descent and its variants. Also, there is no assumption on the parameter space; therefore, the CNN and RNN may be used. It is sample-efficient because several updating epochs are performed between transition sampling which requires PPO less fresh samples to learn the policy network.

Although PPO converges more often to the state-of-the-art methods, there is no convergence guarantee. Additionally, the clipping parameter ϵ should be tuned. PPO imposes itself as the state-of-the-art in Atari games and MuJoco robotic tasks which made it the go-to method in continuous control problems.

Example: PPO [35] has shown comparable or better performance than other policy-based methods, and it has become more attractable (compared to DDPG) to use in continuous tasks. In [42], the authors used PPO to solve the offloading scheduling problem in the vehicular edge computing. The optimization problem of minimizing the long-term cost in terms of energy consumption and delay was formulated as an MDP. Then, a PPO with parameter-shared network architecture combined with a convolutional network was used to learn the value function and the policy.

3.5 Model-Based RL

All the methods we have looked at so far fall under the category of DRL methods called **model-free RL**, where the agent collects data to directly learn a Q-function and/or a policy. Another class of methods we will discuss now is the **model-based RL**, where the actions are selected based on the system environment dynamics, as shown in Figure 3.9. The "model" refers to the system dynamics, precisely the transition distribution $P(s'|s, a)$ that defines the probability of going to state s' from previous state s after taking the action a.

The model can be either known such as in the GO games, or learned using samples of trajectories and transitions from the environment. From the model, we can find a good policy/Q-function that solves the task, and by doing so, we can do the learning without the need to collect new data from the real environment. Hence, the model-based RL methods are well known by their sample efficiency compared to the model-free RL methods, as illustrated in Figure 3.10. At high level, the model-based RL methods that are based on a learned model are based on the process of collection of data, learning the model and later learn the optimal policy, as shown in Figure 3.11. Typically, after learning a policy, we have to go back to collecting more data under the new policy and keep improving the model, and so forth.

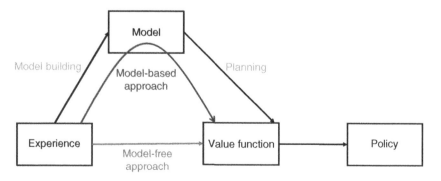

Figure 3.9   Model-free RL vs. model-based RL.

| Gradient-free methods (e.g. NES, CMA) | Fully online methods (e.g. A3C) | Policy gradient methods (e.g. TRPO) | Replay buffer value estimation methods (e.g. DQN, DDPG) | Model-based DRL (e.g. guided policy search) | Model-based shallow RL (e.g. PILCO) |

Better sample efficiency

Figure 3.10 Sample-efficiency of different DRL approaches.

Figure 3.11 Model-based reinforcement learning process.

3.5.1 Vanilla Model-Based RL

The vanilla model-based RL consists of **sampling** from the MDP environment some trajectories using the current policy. Then, using the supervised learning technique, we **train our model** to fit to the data collected. And later on, we **improve the policy** by using the dynamics model with backtracking through time through the learned model, the policy, and the rewards or by using the learned model to build a simulator for model-free RL. We need to iterate over these steps because the model is learned over a limited amount of data so it may be erroneous and imperfect. The initial policy may not generate interesting data and not be as good as we expected. Therefore, we need to iterate over refitting the learned model to new data generated by the updated policy. **Algorithm 3.9** describes the main steps for the Vanilla model-based RL algorithm.

Using model-based RL, we can anticipate data efficiency. In fact, we get a model out of data which might allow significant policy updates than just a policy gradient or value-based methods. And by learning a model, we can use it to solve other tasks

---

Algorithm 3.9 Vanilla model-based RL

1: **for** $k = 1, 2, \ldots$ **do**
2:      Collect data under current policy π and store them into $\mathcal{D} = \{(s, a, s')_i\}$.
3:      Learn dynamics model from previous data by minimizing $\sum_i ||f(s_i, a_i) - s'_i||^2$.
4:      Improve policy by using dynamics model (either by backpropagation through the learned model or by using the learned model as simulator to run RL)
5: **end for**

---

in the same MDP environments because it is not specific to the reward function. The task is defined by the reward function, and the model is independent of the reward function. Hence, we can still reuse the learned model and optimize the new reward function.

However, among the RL methods, the model-based RL methods are the least frequently used ones. One of the reasons is that it is still not mature yet as it could be. It suffers mainly from training instability caused by imperfect learned models. Moreover, it may not achieve the same asymptotic performance as model-free methods.

3.5.2 Robust Model-Based RL: Model-Ensemble TRPO (ME-TRPO)

First, we need to understand the concept of overfitting in model-based RL. Generally, in machine learning techniques such as the supervised learning techniques, overfitting refers to the behavior of the model (e.g. neural network) that performs too well on the training data and poorly on test data. It may happen in model-based RL in the learning of the model that predicts the next state $f(s, a) = s'$. To avoid overfitting, we can use regularization techniques or holdout data. Moreover, there exists a new overfitting challenge in model-based RL. The policy optimization step tends to exploit regions where sufficient data are not available to train the model and this leads to a catastrophic failures. This is called **model bias**. The proposed fix is the **Model-Ensemble Trust Region Policy Optimization** (ME-TRPO). In other words, we suppose that we have a learned simulator that is accurate in some regions and not accurate in other regions. Then, we optimize the policy using the learned simulator. In the regions where the simulator is accurate, the policy behaves optimally, however, for the non-accurate regions it behaves poorly. The learned policy can think that there will be high rewards, which may not be true in the real world. Therefore, if we optimize the policy on such a simulator, it will overfit to the simulator that will be guiding the policy to spaces where the model is not good because it is inaccurate. Hence, the model has some bias that results in policy overfit which we want to avoid.

ME-TRPO [43] was proposed to resolve the issue of model bias in model-based RL methods. In this method, an ensemble of models $(\hat{f}_{\phi_k})_k$ is learned by using supervised learned with sampled data from the environment rather than learning only one model as in vanilla mode-based RL. The ensemble of models defined through neural networks is used to maintain the uncertainty and regularize the learning process to reduce the overfitting. Where there is enough data, the models will agree with their predictions because they are accurate. But in inaccurate regions where there is not enough data, the models will disagree, and thanks to that disagreement we will know that we are outside the region where the simulator is precise and we need to collect data, improve our ensemble models, and

---

Algorithm 3.10 Model Ensemble Trust Region Policy Optimization (ME-TRPO) [45]

1: Initialize a policy π_θ and all models $\hat{f}_{\phi_1}, \hat{f}_{\phi_2}, \dots, \hat{f}_{\phi_K}$.
2: Initialize an empty dataset $\mathcal{B}$.
3: **repeat**
4:      Collect sample from the real system f using π_θ and add them to $\mathcal{D}$.
5:      Train all models using $\mathcal{D}$.
6:      **repeat**
7:          Collect fictitious samples from $\{\hat{f}_{\phi_i}\}_{i=1}^{K}$ using π_θ.
8:          Update the policy using TRPO on the fictitious samples.
9:          Estimate the performances $\hat{\eta}(\theta; \phi_i)$ for $i = 1, \dots, K$.
10:     **until** the performance stop improving.
11: **until** the policy performs well in real environment f.

---

keep repeating. Moreover, the authors proposed the use of the model-free RL algorithm TRPO in place of using backpropagation through time (BPTT) to update the policy. BPTT is known to lead to exploding and vanishing gradient [44], and even with clipping they will get stuck in local optima. Therefore, they use the likelihood method TRPO to estimate the gradient. The ME-TRPO method is summarized in **Algorithm 3.10**.

3.5.3 Adaptive Model-Based RL: Model-Based Meta-Policy Optimization (MB-MPO)

Model-based RL methods are well known for their sample efficiency compared to model-free methods using a learned model in the policy optimization. As discussed before, the model-based RL methods suffer from the model-bias which is an overfitting of the policy to the learned model dynamics disregarding the existing uncertainties in the models. ME-TRPO was proposed to address this issue, and it makes the vanilla model-based RL more robust using an ensemble of models. However, since simulated environments may not be the same as the real environments, the policy that behaves well on learned models may fail in the real environment. Model-Based Meta-Policy Optimization (MB-MPO) tackles this issue by learning an adaptive policy that can quickly adapt to the real world using meta-policy optimization. The key idea of the algorithm is learning an ensemble of model representative of generally how the real world works in addition to learning the adaptive policy. The policy will be able to efficiently work on any new environment that is related to the trained environment. The algorithm is summarized in **Algorithm 3.11**.

Algorithm 3.11 MB-MPO [45]

1: Initialize actor π_θ, the models $\hat{f}_{\phi_1}, \hat{f}_{\phi_2}, \ldots, \hat{f}_{\phi_K}$ and replay buffer $\mathcal{D}$.

2: **repeat**

3:     Add sampled trajectories from the real environment with adapted policies $\pi_{\theta_1'}, \ldots, \pi_{\theta_k'}$ to $\mathcal{D}$.

4:     Train all models $\{\hat{f}_{\phi_k}\}$ using $\mathcal{D}$.

5:     **for** all models $\hat{f}_{\phi_k}$ **do**

6:         Sample imaginary trajectories $\mathcal{T}_k$ from $\hat{f}_{\phi_k}$ using π_θ.

7:         Compute adapted parameters $\theta_k' = \theta + \alpha \nabla_\theta J_k(\theta)$ using trajectories $\mathcal{T}_k$ where:

$$J_k(\theta) = \mathbb{E}_{a_t \sim \pi_\theta(a_t|s_t)} \left[\sum_{t=0}^{H-1} r(s_t, a_t, s_{t+1}) \Big| s_{t+1} = \hat{f}_{\phi_k}(s_t, a_t) \right]. \tag{3.97}$$

8:         Sample imaginary trajectories $\mathcal{T}_k'$ from $\hat{f}_{\phi_k}$ using the adapted policy $\pi_{\theta_k'}$.

9:     **end for**

10:    $\theta \leftarrow \theta - \beta \frac{1}{K} \sum_k \nabla_\theta J_k(\theta_k')$ using the trajectories $\mathcal{T}_k'$.

11: **until** the policy performs well in the real environment

---

3.6 Chapter Summary

We have presented in detail many of the state-of-the-art DRL techniques. We have distinguished two classes of algorithms in RL: the model-free methods where the agent does not have access to the environment dynamics, and the model-based methods in which the agent either has access to the dynamics or learns them. Model-free methods can be divided into two subcategories. The first category, namely, the value-based methods, which learn the value/Q-function from which we can easily select the optimal actions. The second category is the policy-based algorithms such as REINFORCE, DDPG, and TRPO where the agent learns directly the optimal policy. The policy-based algorithms are based on the policy gradient theorem that derives the gradient of the policy to optimize. However, policy-based methods suffer from high sample complexity compared to value-based methods. Moreover, the natural gradients come as an alternative to the classical optimization algorithm (i.e. the SGD) enabling optimization over the probability distribution spaces and it leads to the TRPO method. Again, PPO was proposed later to overcome the limitations of TRPO such as the restriction of using simple neural networks, and it outperforms the other DRL approaches in benchmarking environments to become the go-to method in continuous control problems. Multiple aspects of the DRL-based methods have been investigated to improve the efficiency in solving complex problems. For instance, the exploration technique of the unknown environment may limit the performance and be a bottleneck for the learning process.

References

1 V. Mnih, K. Kavukcuoglu, D. Silver, A. A. Rusu, J. Veness, M. G. Bellemare, A. Graves, M. Riedmiller, A. K. Fidjeland, G. Ostrovski, *et al.*, "Human-level control through deep reinforcement learning," *Nature*, vol. 518, no. 7540, pp. 529–533, 2015.

2 S. Wang, H. Liu, P. H. Gomes, and B. Krishnamachari, "Deep reinforcement learning for dynamic multichannel access in wireless networks," *IEEE Transactions on Cognitive Communications and Networking*, vol. 4, no. 2, pp. 257–265, 2018.

3 X. Wang, Y. Zhang, R. Shen, Y. Xu, and F.-C. Zheng, "DRL-based energy-efficient resource allocation frameworks for uplink NOMA systems," *IEEE Internet of Things Journal*, vol. 7, no. 8, pp. 7279–7294, 2020.

4 H. Van Hasselt, A. Guez, and D. Silver, "Deep reinforcement learning with double Q-learning," in *Proceedings of the AAAI Conference on Artificial Intelligence*, vol. 30, 2016.

5 H. Hasselt, "Double Q-learning," *Advances in neural information processing systems 23*, 2010.

6 Y. Al-Eryani, M. Akrout, and E. Hossain, "Multiple access in cell-free networks: Outage performance, dynamic clustering, and deep reinforcement learning-based design," *IEEE Journal on Selected Areas in Communications*, vol. 39, no. 4, pp. 1028–1042, 2020.

7 T. Schaul, J. Quan, I. Antonoglou, and D. Silver, "Prioritized experience replay," *arXiv preprint arXiv:1511.05952*, 2015.

8 L. C. Baird III, "Advantage updating," tech. rep., WRIGHT LAB WRIGHT-PATTERSON AFB OH, 1993.

9 Z. Wang, T. Schaul, M. Hessel, H. Hasselt, M. Lanctot, and N. Freitas, "Dueling network architectures for deep reinforcement learning," in *International Conference on Machine Learning*, pp. 1995–2003, PMLR, 2016.

10 N. Zhao, Y.-C. Liang, D. Niyato, Y. Pei, M. Wu, and Y. Jiang, "Deep reinforcement learning for user association and resource allocation in heterogeneous cellular networks," *IEEE Transactions on Wireless Communications*, vol. 18, no. 11, pp. 5141–5152, 2019.

11 R. J. Williams, "Simple statistical gradient-following algorithms for connectionist reinforcement learning," *Machine Learning*, vol. 8, no. 3, pp. 229–256, 1992.

12 R. S. Sutton, D. McAllester, S. Singh, and Y. Mansour, "Policy gradient methods for reinforcement learning with function approximation," *Advances in neural information processing systems 12*, 1999.

13 Y. Yang, Y. Li, K. Li, S. Zhao, R. Chen, J. Wang, and S. Ci, "DECCO: Deep-learning enabled coverage and capacity optimization for massive MIMO systems," *IEEE Access*, vol. 6, pp. 23361–23371, 2018.

14 J. Vitay, "1 introduction."

15 S. Gu, E. Holly, T. Lillicrap, and S. Levine, "Deep reinforcement learning for robotic manipulation with asynchronous off-policy updates," in *2017 IEEE International Conference on Robotics and Automation (ICRA)*, pp. 3389–3396, IEEE, 2017.

16 J. Heydari, V. Ganapathy, and M. Shah, "Dynamic task offloading in multi-agent mobile edge computing networks," in *2019 IEEE Global Communications Conference (GLOBECOM)*, pp. 1–6, IEEE, 2019.

17 V. Mnih, A. P. Badia, M. Mirza, A. Graves, T. Lillicrap, T. Harley, D. Silver, and K. Kavukcuoglu, "Asynchronous methods for deep reinforcement learning," in *International Conference on Machine Learning*, pp. 1928–1937, PMLR, 2016.

18 B. Recht, C. Re, S. Wright, and F. Niu, "Hogwild!: A lock-free approach to parallelizing stochastic gradient descent," *Advances in neural information processing systems 24*, 2011.

19 C. J. C. H. Watkins, "Learning from delayed rewards," 1989.

20 R. J. Williams and J. Peng, "Function optimization using connectionist reinforcement learning algorithms," *Connection Science*, vol. 3, no. 3, pp. 241–268, 1991.

21 Z. Wang, L. Li, Y. Xu, H. Tian, and S. Cui, "Handover optimization via asynchronous multi-user deep reinforcement learning," in *2018 IEEE International Conference on Communications (ICC)*, pp. 1–6, 2018.

22 J. Schulman, P. Moritz, S. Levine, M. Jordan, and P. Abbeel, "High-dimensional continuous control using generalized advantage estimation," *arXiv preprint arXiv:1506.02438*, 2015.

23 Y. Jang, S. M. Raza, H. Choo, and M. Kim, "UAVs handover decision using deep reinforcement learning," in *2022 16th International Conference on Ubiquitous Information Management and Communication (IMCOM)*, pp. 1–4, IEEE, 2022.

24 D. Silver, G. Lever, N. Heess, T. Degris, D. Wierstra, and M. Riedmiller, "Deterministic policy gradient algorithms," in *International Conference on Machine Learning*, pp. 387–395, PMLR, 2014.

25 T. P. Lillicrap, J. J. Hunt, A. Pritzel, N. Heess, T. Erez, Y. Tassa, D. Silver, and D. Wierstra, "Continuous control with deep reinforcement learning," *arXiv preprint arXiv:1509.02971*, 2015.

26 G. E. Uhlenbeck and L. S. Ornstein, "On the theory of the Brownian motion," *Physical Review*, vol. 36, no. 5, p. 823, 1930.

27 S. Ioffe and C. S. B. Normalization, "Accelerating deep network training by reducing internal covariate shift," *arXiv preprint arXiv:1502.03167*, 2014.

28 F. Fredj, Y. Al-Eryani, S. Maghsudi, M. Akrout, and E. Hossain, "Distributed beamforming techniques for cell-free wireless networks using deep reinforcement learning," *IEEE Transactions on Cognitive Communications and Networking*, vol. 8, no. 2, pp. 1186–1201, 2022.

29 G. Barth-Maron, M. W. Hoffman, D. Budden, W. Dabney, D. Horgan, D. Tb, A. Muldal, N. Heess, and T. Lillicrap, "Distributed distributional deterministic policy gradients," *arXiv preprint arXiv:1804.08617*, 2018.

30 M. G. Bellemare, W. Dabney, and R. Munos, "A distributional perspective on reinforcement learning," in *International Conference on Machine Learning*, pp. 449–458, PMLR, 2017.

31 S.-I. Amari, "Natural gradient works efficiently in learning," *Neural Computation*, vol. 10, no. 2, pp. 251–276, 1998.

32 S. M. Kakade, "A natural policy gradient," *Advances in neural information processing systems 14*, 2001.

33 J. Peters and S. Schaal, "Reinforcement learning of motor skills with policy gradients," *Neural Networks*, vol. 21, no. 4, pp. 682–697, 2008.

34 J. Schulman, S. Levine, P. Abbeel, M. Jordan, and P. Moritz, "Trust region policy optimization," in *International Conference on Machine Learning*, pp. 1889–1897, PMLR, 2015.

35 J. Schulman, F. Wolski, P. Dhariwal, A. Radford, and O. Klimov, "Proximal policy optimization algorithms," *arXiv preprint arXiv:1707.06347*, 2017.

36 M. Arjovsky, S. Chintala, and L. Bottou, "Wasserstein generative adversarial networks," in *International Conference on Machine Learning*, pp. 214–223, PMLR, 2017.

37 I. Goodfellow, J. Pouget-Abadie, M. Mirza, B. Xu, D. Warde-Farley, S. Ozair, A. Courville, and Y. Bengio, "Generative adversarial nets," *Advances in neural information processing systems 27*, 2014.

38 R. Pascanu and Y. Bengio, "Revisiting natural gradient for deep networks," *arXiv preprint arXiv:1301.3584*, 2013.

39 S. Kakade and J. Langford, "Approximately optimal approximate reinforcement learning," in *Proceedings of the 19th International Conference on Machine Learning*, Citeseer, 2002.

40 A. Feriani and E. Hossain, "Single and multi-agent deep reinforcement learning for AI-enabled wireless networks: A tutorial," *IEEE Communications Surveys & Tutorials*, vol. 23, no. 2, pp. 1226–1252, 2021.

41 A. Valadarsky, M. Schapira, D. Shahaf, and A. Tamar, "A machine learning approach to routing," *arXiv preprint arXiv:1708.03074*, 2017.

42 W. Zhan, C. Luo, J. Wang, C. Wang, G. Min, H. Duan, and Q. Zhu, "Deep-reinforcement-learning-based offloading scheduling for vehicular edge computing," *IEEE Internet of Things Journal*, vol. 7, no. 6, pp. 5449–5465, 2020.

43 T. Kurutach, I. Clavera, Y. Duan, A. Tamar, and P. Abbeel, "Model-ensemble trust-region policy optimization," *arXiv preprint arXiv:1802.10592*, 2018.

44 Y. Bengio, P. Simard, and P. Frasconi, "Learning long-term dependencies with gradient descent is difficult," *IEEE Transactions on Neural Networks*, vol. 5, no. 2, pp. 157–166, 1994.

45 I. Clavera, J. Rothfuss, J. Schulman, Y. Fujita, T. Asfour, and P. Abbeel, "Model-based reinforcement learning via meta-policy optimization," in *Conference on Robot Learning*, pp. 617–629, PMLR, 2018.

4

A Case Study and Detailed Implementation

4.1 System Model and Problem Formulation

Wireless communications are extremely vulnerable to jamming attacks due to the broadcast nature of transmission, especially in low-power wireless networks such as IoT and wireless sensor networks [1]. In particular, a jammer can disrupt or even take down the legitimate communications between the transmitter and the receiver by sending strong interference signals to the target channel, thereby significantly reducing the effective signal-to-interference-plus-noise ratio (SINR) at the receiver [2, 3]. More seriously, jamming attacks can be easily performed with common circuits and device [4, 5].

There are several ways to prevent and reduce the effects of jamming attacks [6]. One simple and common approach is to adjust the transmit power of the legitimate transmitter. The transmitter can transmit at low power to make it harder for the jammer to detect the signals, or at high power to overpower the jamming signals at the receiver. While the former is only feasible in dealing with reactive jammers and significantly reduces the transmission efficiency, the latter requires more power and may not be effective when the jammer attacks the channel with very high-power levels. Another solution is frequency-hopping spread spectrum (FHSS) transmission [6, 7]. With FHSS transmission, the transmitter can switch to another communication channel to transmit its data when the current channel is under attack. Unfortunately, this solution requires a set of available communication channels as well as a predefined switching algorithm. Moreover, if the jammer has enough power to attack multiple channels (or even all available channels) at the same time, the FHSS mechanism is not effective. The rate adaptation (RA) is also a common technique in the literature to cope with jamming attacks. This technique allows a legitimate transmitter to reduce its transmission rate based on the jamming power. Nevertheless, this solution is either power-inefficient or not viable for low-power or hardware-constrained devices (e.g. in Internet-of-Things [IoT] applications) [1].

Deep Reinforcement Learning for Wireless Communications and Networking:
Theory, Applications, and Implementation, First Edition.
Dinh Thai Hoang, Nguyen Van Huynh, Diep N. Nguyen, Ekram Hossain, and Dusit Niyato.
© 2023 The Institute of Electrical and Electronics Engineers, Inc. Published 2023 by John Wiley & Sons, Inc.

To address all the limitations of conventional anti-jamming approaches, we introduce a method that allows the transmitter to communicate even under jamming attacks. In particular, the transmitter is equipped with energy harvesting and ambient backscatter capabilities. The energy harvesting circuit allows the transmitter to harvest energy from the jamming signals to support its operations, while the ambient backscatter capability allows the transmitter to use the strong jamming signals to send its data to the receiver even under attack. This enables the transmitter to not only evade the jamming attacks but also make use of the jamming signals for its own transmissions. It is important to note that the jammer's attack strategy is not known in advance to the transmitter, and the dynamic and uncertain nature of the wireless environment makes it difficult to determine the optimal anti-jamming strategy. In this context, reinforcement learning can be an effective tool because it allows an agent to interact with the system, observe the outcomes, and act accordingly, enabling it to learn the jammer's attack policy and the system's characteristics in order to identify the optimal strategy for the transmitter.

In the following, we will first provide the system model in detail. Then, a Markov decision process (MDP) model for the considered problem will be formulated. After that, we will provide the detailed implementation of a deep reinforcement learning (DRL) algorithm to solve the formulated MDP. Finally, we will discuss the performance of the proposed DRL method and the effects of learning parameters on its learning process.

4.1.1 System Model and Assumptions

We consider a wireless network consisting of a receiver, a transmitter, and a jammer as shown in Figure 4.1. The transmitter is equipped with an energy harvesting circuit and a backscatter circuit [8]. The transmitter thus can either harvest energy and use the harvested energy to actively transmit its data, i.e. harvest-then-transmit (HTT) mode, or backscatter data on the jamming signals [8], i.e. backscatter mode.

4.1.1.1 Jamming Model

In the system under consideration, the jammer attacks the channel to degrade the effective SINR at the receiver [9]. The SINR can be calculated as $\theta = \frac{P_R}{\phi P_J + \rho^2}$, [4, 9], in which P_J is the jamming power, P_R denotes the received power at the receiver, and ρ^2 is the variance of additive white Gaussian noise. Denote $0 \leq \phi \leq 1$ is the channel attenuation factor, the jamming power received at the receiver can be formulated as ϕP_J. We then denote $\mathbf{P}_J = \{P_0^J, \ldots, P_n^J, \ldots, P_N^J\}$ as a vector of discrete jamming power levels from P_0^J to P_N^J. At each time slot, the jammer selects a given jamming power level P_n^J with a given probability x_n as long as its average

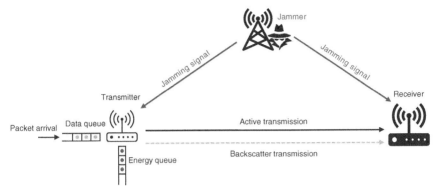

Figure 4.1 System model. Adapted from Van Huynh et al. [2].

power constraint is satisfied. Specifically, let $\mathbf{J}_s$ denote the strategy space of the jammer and $\mathbb{X}$ as the attack probability vector, then we have

$$\mathbf{J}_s = \left\{ \mathbf{x} = (x_0, \dots, x_n, \dots, x_N), \sum_{i=0}^{N} x_n = 1 \right\}. \tag{4.1}$$

4.1.1.2 System Operation

When the system is not being attacked by the jammer, the transmitter has the option to either remain inactive or transmit a maximum of $\hat{d}_t$ packets to the receiver using normal transmission methods, which requires e_t units of energy per packet. On the other hand, if the system is being attacked by the jammer, the transmitter can perform rate adaptation, harvest energy from the jamming signals, or backscatter the jamming signals to send its data to the receiver.

Specifically, the transmitter can still transmit data by adapting its transmission rate to the jamming power. The set of available transmission rates for the transmitter in this situation is denoted as $\mathbf{r} = r_1, \dots, r_m, \dots, r_M$. At the jamming power level P_n^J, the transmitter can transmit at the highest rate of r_m, but any packets transmitted at rates $r_{m+1}, \dots, r_M$ will be lost. The transmitter can send a maximum of $\hat{d}_m^r$ packets at each rate r_m. It is important to note that the transmitter can only determine if the jammer is active or inactive, but not the exact jamming power level. The proposed reinforcement-learning algorithm allows the transmitter to learn about the jamming power level and the corresponding likelihood, and then adapt its anti-jamming strategy to maximize the average long-term throughput.

Under jamming attacks, the transmitter can also choose to harvest energy from the jamming signals. The harvested energy is then used for active transmissions. We denote e_n^J as the amount of energy harvested by the transmitter when the jammer attacks the system with the power level of P_n^J. In addition, the transmitter can use the ambient backscatter circuit to backscatter information right on the jamming signals. We denote $\hat{d}_n^J$ as the maximum number of packets backscattered by the transmitter when the jamming power is P_n^J. It is worth noting that the backscatter rate is defined by its circuits and may not change when operating [8]. As such, in this example, we assume that if the transmitter chooses to backscatter the jamming signals, it will backscatter $b^\dagger$ packets. If $b^\dagger > \hat{d}_n^J$, $(b^\dagger - \hat{d}_n^J)$ packets will be lost during the backscattering process. We denote $\mathbf{e} = \{e_0^J, \dots, e_N^J\}$ and $\hat{\mathbf{d}} = \{\hat{d}_0^J, \dots, \hat{d}_N^J\}$ as the amount of harvested energy and the backscattered packet vectors of the transmitter, respectively. It is worth noting that, the stronger the jamming signal, the more number of packets the transmitter can backscatter, and the more energy the transmitter can harvest [2]. These relationships can be either linear or nonlinear, and our proposed framework with the following analysis is applicable to both. We denote D and E as the maximum size of the data queue and the energy queue at the transmitter, respectively. The data arrival process is assumed to follow a Poisson distribution with rate λ. If a packet arrives at the transmitter when the data queue is full, it will be dropped.

4.1.2 Problem Formulation

To consider the dynamic and unpredictable nature of the jammer, we express the optimization problem of the system as an MDP. This approach enables the transmitter to choose the optimal action at a given state in order to maximize its long-term average reward, which in this case is the average throughput of the system. The MDP is defined by a tuple $\langle S, A, r \rangle$, where S is the state space of the system, A is the action space, and r is the immediate reward function of the system.

4.1.2.1 State Space

The system state space can be defined as the combination of the states of the jammer, the data queue, and the energy queue as follows:

$$S = \{(j, d, e) : j \in \{0, 1\}; d \in \{0, \dots, D\}; e \in \{0, \dots, E\}\}, \tag{4.2}$$

where j is the state of the jammer, i.e. $j = 0$ if the jammer is idle and $j = 1$ if the jammer attacks the system. d and e represent the number of packets in the data queue and the number of energy units in the energy queue. The state of the system is then defined as a composite variable $s = (j, d, e) \in S$. It is worth mentioning that, to highlight the effect of the backscattering and energy harvesting from the jamming signals, we assume that the transmitter has no energy at the beginning (i.e. the worst case in which the transmitter has no energy and being attacked). If the transmitter is powered with a battery or if the energy is not of concern, our model can still be applied by setting the transmitter with a nonzero or infinite initial energy level, respectively.

4.1.2.2 Action Space

The transmitter has a total of $(M + 4)$ actions to choose from, which include *remaining idle, actively transmitting data, harvesting energy from the jamming signals, backscattering the jamming signals,* and *adapting its transmission rate* to one of M rates using the RA technique when the jammer is attacking the system. Then, the action space can be defined by $\mathcal{A} \triangleq \{a : a \in \{1, \dots, M+4\}\}$, where

$$
a = \begin{cases}
1, & \text{stay idle,} \\
2, & \text{actively transmit data,} \\
3, & \text{harvest energy from the jamming signals,} \\
4, & \text{backscatter data using the jamming signals,} \\
4 + m, & \text{adapt the transmission rate to} \\
& \quad \text{rate } r_m \text{ with } m \in \{1, \dots, M\}.
\end{cases}
\tag{4.3}
$$

The transmitter's actions are illustrated in a flowchart in Figure 4.2. In particular, the transmitter always has the option to stay idle. If the jammer is inactive, and there are data available in the data queue and sufficient energy in the energy queue, the transmitter can choose to actively transmit data to the receiver. On the other hand, if the jammer is attacking the channel, the transmitter can harvest

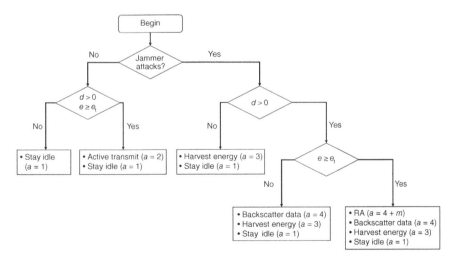

Figure 4.2 Flowchart to express the actions of the transmitter.

energy from the jamming signals or, if there are data in the data queue, it can backscatter data using the jamming signals. Additionally, if there is enough energy in the energy queue, the transmitter can use the RA technique to transmit data to the receiver.

4.1.2.3 Immediate Reward

In this example, we defined the immediate reward after performing action a at state s as the number of packets that are successfully transmitted to the receiver. As such, the immediate reward $r(s, a)$ can be expressed as follows:

$$
r(s, a) = \begin{cases} d_t, & (j = 0, d > 0, e \geq e_t, a = 2; 0 < d_t \leq \hat{d}_t), \\ d_n^J, & (j = 1, d > 0, a = 4; 0 < d_n^J \leq \hat{d}_n^J), \\ d_m^r, & (j = 1, d > 0, e \geq e_t, a = 4 + m; 0 < d_m^r \leq \hat{d}_m^r), \\ 0, & \text{otherwise.} \end{cases} \tag{4.4}
$$

In the above, when the jammer is idle (i.e. $j = 0$), the transmitter has data in the data queue and has enough energy in the energy queue, it can choose to actively transmit (i.e. $a = 2$) $0 < d_t \leq \hat{d}_t$ packets to the receiver. Otherwise, when the jammer attacks the system, the transmitter can backscatter the jamming signals (i.e. $a = 4$) to backscatter $0 < d_n^J \leq \hat{d}_n^J$) packets if it has data in the data queue. In addition, the transmitter can also choose to adapt its rate to r_m (i.e. $a = 4 + m$) to actively transmit $0 < d_m^r \leq \hat{d}_m^r$ packets to the receiver if it has data in the data queue. Finally, the immediate reward is equal to 0 if the transmitter cannot successfully transmit any packet to the receiver.

4.1.2.4 Optimization Formulation

Our goal is to find the optimal policy, denoted by π^*, that maximizes the average long-term throughput of the system. This policy, π^*, is a function that maps a given state, including the jammer attack state, the data queue state, and the energy queue state, to the optimal action. The optimization problem can be stated as follows:

$$
\max_{\pi} \mathcal{R}(\pi) = \lim_{T \to \infty} \frac{1}{T} \sum_{k=1}^{T} \mathbb{E}\left(r_k(s_k, \pi(s_k))\right), \tag{4.5}
$$

where $\mathcal{R}(\pi)$ denotes the average throughput of the transmitter under policy π and $r_k(s_k, \pi(s_k))$ is the immediate reward at time slot k after performing action a_k at state s_k, i.e. $a_k = \pi(s_k)$. It's important to note that the formulated MDP is irreducible, which means that from any given state, it is always possible to reach any other state after a certain number of steps, and there is no terminal state. This ensures that for any policy π, the average throughput $\mathcal{R}(\pi)$ is well defined and does not depend on the initial state [10].

4.2   Implementation and Environment Settings

In this section, we present in detail the implementations of several reinforcement learning algorithms to solve the optimization problem defined in (4.5). In particular, we first present a guideline to install TensorFlow and required packages to run the provided codes by using Anaconda. Then, we introduce the codes for the conventional Q-learning algorithm. After that, the implementation of the deep Q-learning algorithm is provided.

4.2.1   Install TensorFlow with Anaconda

There are several methods to install TensorFlow such as using Python's pip command or using TensorFlow Docker images. However, these approaches are either complicated or require different command lines for different operating systems. In this tutorial, we use Anaconda as an alternative approach to install TensorFlow. TensorFlow with Anaconda is supported on 64-bit Windows 7 or later, 64-bit Ubuntu Linux 14.04 or later, 64-bit CentOS Linux 6 or later, and macOS 10.10 or later. The following instructions are the same for all these operating systems.

1. Download and install Anaconda from their website at https://www.anaconda .com/products/distribution.
2. On Linux and macOS, open a terminal window and use the default bash shell to open Anaconda. Please refer to Anaconda's website for more detail.
3. To install the current release of TensorFlow that runs on CPU only, use the following commands:

```
1  conda create -n tf tensorflow
2  conda activate tf
```

The first command is used to `create` a new virtual conda environment, named `tf`, with TensorFlow package installed. The second command is used to activate the `tf` environment. After that, TensorFlow is ready to use in the `tf` environment. With Anaconda, we can create as many virtual environments with different packages installed as we want.

4. To install TensorFlow with graphics processing unit (GPU) supported, use the following commands:

```
1  conda create -n tf-gpu tensorflow-gpu
2  conda activate tf-gpu
```

In the above commands, `tensorflow-gpu` is the TensorFlow package that supports GPU computations.

It is worth noting that the provided codes of Q-learning and deep Q-learning below are checked and work well with Python 3.9, TensorFlow 2.9.1, Keras 2.9, and NumPy 1.22.4.[1] With Anaconda, we can easily install specific versions of TensorFlow, Python, Numpy, and Keras as follows:

```
1  conda create -n tf-drl-book tensorflow=2.9.1 python=3.9
       keras=2.9 numpy=1.22.4
2  conda activate tf-drl-book
```

After creating the conda environment with TensorFlow and related packages, the readers can run Python codes by running the below command, with `script_name` is the name of the Python script.

```
1  python script_name.py
```

4.2.2 Q-Learning

In this section, we provide the detailed code for the Q-learning algorithm. Before moving to the main codes, let us define the parameters that we will be using as shown below.

parameters.py

```
1  nu = 0.1
2  arrival_rate = 3
3  nu_p = [0.6, 0.2, 0.2]
4  d_t = 4
5  e_t = 1
6  d_bj_arr = [1, 2, 3]
7  e_hj_arr = [1, 2, 3]
8  dt_ra_arr = [2, 1, 0]
9  d_queue_size = 10
10 e_queue_size = 10
11 b_dagger = 3
12 num_actions = 7
13 num_states = 2 * (d_queue_size + 1) * (e_queue_size + 1)
14
15 learning_rate_Q = 0.1
16 gamma_Q = 0.9
17 learning_rate_deepQ = 0.0001
18 gamma_deepQ = 0.99
19
20 num_features = 3
21 memory_size = 10000
22 batch_size = 32
23 update_target_network = 5000
24 step = 1000 # print reward after every step
25 T = 1000000 # number of training iterations
```

In the code above, "nu" is the probability that the jammer is idle, "`arrival_rate`" is the mean rate of the data arrival process, "nu_p" is

---

1 Other versions may also work. However, we recommend installing the mentioned versions.

the probabilities that the jammer attacks with power levels 1, 2, and 3. "d_t" is the number of packets per active transmission, "e_t" is the number of energy units required for actively transmitting a packet, "d_bj_arr" is the number of packets that can be backscattered corresponding to jamming power levels 1, 2, and 3. "e_hj_arr" is the number of energy units that can be harvested corresponding to jamming power levels 1, 2, and 3, "dt_ra_arr" is the number of packets that can be transmitted by using the RA technique corresponding to jamming power levels 1, 2, and 3. "d_queue_size" and "e_queue_size" present the capacity of the data queue and the energy queue, respectively. "b_dagger" is the fixed number of packets that can be backscattered. "num_actions" is the number of actions in the action space. In this example, the jammer attacks the system with three power levels, as such the total number of actions is 7. "num_states" is the total number of states in the state space.

"learning_rate_Q" and "gamma_Q" represent the learning rate and discount factor for the Q-learning algorithm, respectively. In this example, we choose the common values for these parameters, i.e. 0.1 and 0.9, respectively. "learning_rate_deepQ" and "gamma_deepQ" represent the learning rate and discount factor for the deep Q-learning algorithm, respectively. These parameters will be varied later in Section 4.3 to show their effects on the learning process of the deep Q-learning algorithm. "num_features" denotes the number of features in the state space, i.e. jammer state, data queue state, and energy queue state. "memory_size," "batch_size," and "update_target_network" represent the capacity of the memory pool, the batch size, and the frequency of updating the target Q-network.

4.2.2.1 Codes for the Environment

As we know, in a reinforcement-learning algorithm, there are two entities, including the agent and the environment. In the following, we present the code for the environment, i.e. environment class. First, we include the necessary packages and initialize the jammer state, the data queue state, and the energy queue state.

environment.py

```
1  from parameters import *
2  import numpy as np
3  import random
4  from scipy.stats import poisson
5
6  class environment:
7    def __init__(self):
8      self.jammer_state = 0
9      self.data_state = 0
10     self.energy_state = 0
```

Next, we will need to construct a get state function to help agent obtain the current state of the environment as follows:

```
environment.py
1  def get_state(self):
2     count = 0
3     state = 0
4     for jammer in range(0, 2):
5        for data in range(0, d_queue_size + 1):
6           for energy in range(0, e_queue_size + 1):
7              if self.jammer_state == jammer and self.
                  data_state == data and self.energy_state
                  == energy:
8                 state = count
9              count += 1
10    return state
```

The Q-learning algorithm deploys a Q-table to save the Q-values of all state-action pairs. As such, in the above, we need to convert the current state from a combination of jammer state, data queue state, and energy queue state to a decimal number. Next, the transmitter will determine a set of possible actions given the current system state as in the function below.

```
environment.py
1  def get_possible_action(self):
2     list_actions = [0]  # stay idle
3     if self.jammer_state == 0 and self.data_state > 0
         and self.energy_state >= e_t:
4        list_actions.append(1)  # active transmit
5     if self.jammer_state == 1:
6        list_actions.append(2)  # harvest energy
7        if self.data_state > 0:
8           list_actions.append(3)   # backscatter
9           if self.energy_state >= e_t:
10             list_actions.append(4)   # RA1
11             list_actions.append(5)   # RA2
12             list_actions.append(6)   # RA3
13    return list_actions
```

As mentioned above, the transmitter can always choose to remain idle. If the jammer is inactive, the transmitter buffer is not empty, and there is sufficient energy in the energy queue, the transmitter can actively transmit data. If the jammer is active, the transmitter can harvest energy from the jamming signals or backscatter data if it has data to send to the receiver. Additionally, the transmitter can use rate adaptation if there is enough energy in the energy queue. The following function is used to calculate the immediate reward after performing an action.

```
environment.py
1  def calculate_reward(self, action):
2      reward = 0
3      loss = 0
4      if action == 0:
5          reward = 0
6      elif action == 1:  # actively transmit
7          reward = self.active_transmit(d_t)
8      elif action == 2:  # harvest energy
9          reward = random.choices(e_hj_arr, nu_p, k=1)[0]
10     elif action == 3:  # backscatter
11         d_bj = random.choices(d_bj_arr, nu_p, k=1)[0]
12
13         if self.data_state >= b_dagger:
14             max_rate = b_dagger
15         else:
16             max_rate = self.data_state
17
18         if self.data_state > d_bj:
19             reward = d_bj
20         else:
21             reward = self.data_state
22
23         if max_rate > reward:
24             loss = max_rate - reward
25
26     elif action == 4:  # RA1
27         max_ra = random.choices(dt_ra_arr, nu_p, k=1)[0]
28         reward = self.active_transmit(dt_ra_arr[0])
29
30         if dt_ra_arr[0] > max_ra:  # selected rate higher
                                        than the successful rate
31             loss = reward
32             reward = 0
33
34     elif action == 5:  # RA2
35         max_ra = random.choices(dt_ra_arr, nu_p, k=1)[0]
36         reward = self.active_transmit(dt_ra_arr[1])
37
38         if dt_ra_arr[1] > max_ra:  # selected rate higher
                                        than the successful rate
39             loss = reward
40             reward = 0
41     elif action == 6:  # RA3
42         reward = 0
43
44     return reward, loss
```

The transmitter has several options for actions: staying idle, actively transmitting data, harvesting energy from jamming signals, backscattering data, or performing rate adaptation. If the transmitter stays idle, the immediate reward will be zero. If the transmitter chooses to actively transmit data, the reward will depend on the current number of packets in the data queue and the number of energy units in the energy queue (refer to the "active_transmit" below for more details). If the energy in the queue is sufficient, the transmitter can actively transmit data to the receiver. If the transmitter chooses to harvest energy, the number of harvested energy units will be randomly determined in "e_hj_arr" based on certain attack power level probabilities "nu_p." If the transmitter chooses to backscatter data, it will first determine the maximum number of backscattered packets based on the jammer's transmit power. As the transmitter always backscatters data at a fixed rate, i.e. always backscatter "b_dagger" packets, there will be packet loss if "b_dagger" is higher than the number of backscattered packets. If the transmitter chooses to perform rate adaptation, the maximum number of packets that can be transmitted will be obtained from the "dt_ra_arr" given the attack power level probabilities, and the reward will be calculated based on the current number of packets in the data queue and the number of energy units in the energy queue.

---

environment.py

```
1   def active_transmit(self, maximum_transmit_packets):
2       num_transmitted = 0
3       if 0 < self.data_state < maximum_transmit_packets:
4           if self.energy_state <= e_t * self.data_state:
5               num_transmitted = self.data_state
6           elif self.energy_state <= e_t:
7               num_transmitted = self.energy_state // e_t
8       else:
9           if self.energy_state <= e_t *
                   maximum_transmit_packets:
10              num_transmitted = maximum_transmit_packets
11          elif self.energy_state <= e_t:
12              num_transmitted = self.energy_state // e_t
13      return num_transmitted
```

After calculating the immediate reward, the system will be updated by the "perform_action" function below.

---

environment.py

```
1   def perform_action(self, action):
2       reward, loss = self.calculate_reward(action)
3       if action == 1:
```

```
4     self.data_state -= reward
5     self.energy_state -= reward * e_t
6   elif action == 2:
7     if self.energy_state < e_queue_size:
8       self.energy_state += reward
9     if self.energy_state > e_queue_size:
10      self.energy_state = e_queue_size
11    reward = 0
12  elif action == 3:    # when perform backscatter,
      always backscatter 3 packages
13    if self.data_state >= b_dagger:
14      max_rate = b_dagger
15    else:
16      max_rate = self.data_state
17
18    self.data_state -= max_rate
19  elif action == 4 or action == 5:
20    if reward > 0:
21      self.data_state -= reward
22      self.energy_state -= reward * e_t
23    else:
24      self.data_state -= loss
25      self.energy_state -= loss * e_t
26
27  # data arrival
28  data_arrive_l = poisson.rvs(mu=arrival_rate, size=1)
29  data_arrive = data_arrive_l[0]
30  self.data_state += data_arrive
31  if self.data_state > d_queue_size:
32    self.data_state = d_queue_size
33
34  # jammer state
35  if self.jammer_state == 0:
36    if np.random.random() <= 1 - nu:
37      self.jammer_state = 1
38  else:
39    if np.random.random() <= nu:
40      self.jammer_state = 0
41
42  next_state = self.get_state()
43  return reward, next_state
```

In the above function, the number of data packets in the data queue and the number of energy units in the energy queue will be updated according to the chosen action of the transmitter. Note that the immediate reward is the number of packets that are successfully transmitted to the receiver. As such, the reward of harvesting energy is set to zero. As the harvested energy can be used to transmit data, reinforcement learning can learn and take actions related to energy harvesting given the current conditions of the system. For example, if the energy queue is full, the transmitter can choose not to harvest energy from the jamming signals. When the jammer is likely to attack the system with high-power levels, the algorithm can learn and only take the backscattering action as the active transmission (which uses the harvested energy) will be likely to be failed. After performing the action, the data arrival is simulated following the Poisson distribution. The state of the jammer, i.e. idle or active, will also be generated. Finally, the next system state and the immediate reward will be returned to the transmitter for learning.

4.2.2.2 Codes for the Agent

In this section, we provide the codes for the agent of the Q-learning algorithm. First, we will need to import the required packets and classes. Then, the environment and the Q-table are initialized as shown in the codes below.

---

q_learning_agent.py

```
1   from environment import environment
2   import numpy as np
3   from parameters import *
4
5   class q_learning_agent:
6     def __init__(self):
7       self.env = environment()
8       self.q_matrix = np.zeros((num_states,num_actions))
```

Then, the learning process of the Q-learning algorithm is presented in the "learning" function below. First, we initialize the epsilon value for the epsilon-greedy method. In particular, the epsilon value will be gradually decreased from 1 to 0.01. In this way, the learning process will be likely to perform the exploration process when it starts running, and then gradually move to the exploitation process in later learning rounds. At each learning round, the agent first gets the current system state by calling the function "get_state" from the environment class. Then, the list of possible actions given the current state is obtained. The epsilon-greedy method is then performed by randomly taking an action from the list of possible actions with the probability of epsilon or taking an action that has the highest Q-value in the Q-table with the probability of (1 − *epsilon*). The immediate reward and next state after performing the selected action will be obtained by calling function "perform_action" from the environment class. Then, the highest Q-value of the next state is obtained to calculate the Bellman equation to update the Q-table. The learning process then moves to the next learning steps until it is converged or terminated.

q_learning_agent.py

```
1  def learning(self):
2    epsilon = 1
3    decay = 0.9999
4    min_epsilon = 0.01
5    action = 0
6    total_reward = 0
7    for i in range(T):
8      current_state = self.env.get_state()
9      list_possible_actions = self.env.
         get_possible_action()
10     max_q = -float("inf")
11     if np.random.random() <= epsilon:
12       action = np.random.choice(list_possible_actions)
13     else:
14       for action_t in list_possible_actions:
15         if self.q_matrix[current_state][action_t] >
             max_q:
16           max_q = self.q_matrix[current_state][
               action_t]
17           action = action_t
18
19     reward, next_state = self.env.perform_action(
         action)
20     total_reward += reward
21     list_possible_next_actions = self.env.
         get_possible_action()
22     max_q = -float("inf")
23     for action_n in list_possible_next_actions:
24       if self.q_matrix[next_state][action_n] >= max_q:
25         max_q = self.q_matrix[next_state][action_n]
26
27     data = (1-learning_rate_Q)*self.q_matrix[
         current_state][action] + learning_rate_Q*(
         reward+discount_factor_Q*max_q)
28     self.q_matrix[current_state][action] = data
29     temp = epsilon * decay
30     epsilon = max(min_epsilon, temp)
31     if (i+1) % step == 0:
32       print("Iteration " + str(i + 1) + " reward: " +
           str(total_reward / (i + 1)))
```

4.2.3 Deep Q-Learning

The Q-learning algorithm is designed to find the optimal policy with a probability of 1, but in practice, it may not converge quickly, especially in complex systems with large state and action spaces. Additionally, in many situations, the state and/or action space may be continuous, making it impossible for the Q-learning algorithm to construct a Q-table. The deep Q-learning algorithm can efficiently address these limitations. The following information explains the implementation of a deep Q-learning algorithm.

Similar to the Q-learning algorithm, there are also two entities in the deep Q-learning algorithm, i.e. the environment and the deep Q-learning agent. The codes of the environment class are mostly the same as those of the Q-learning algorithm provided above. There are two functions that we need to modify is "get_state" and "perform_action." In particular, the "get_state" function is changed to the "get_state_deep" as follows:

---

environment.py

```
1  def get_state_deep(self):
2      return np.array([self.jammer_state, self.data_state,
           self.energy_state])
```

As can be seen in the "get_state_deep" function, the state space is not required to be discrete as in the Q-learning algorithm. Instead, the states of the jammer, the data queue, and the energy queue are now directly fed to the deep neural network. In the "perform_action," we will also need to change "next_state = self.get_state()" to "next_state = self.get_state_deep()." The modified function will be named "perform_action_deep."

Next, we will construct the deep Q-learning agent. First, we import the required packets and the classes as follows:

---

deep_q_learning_agent.py

```
1  from environment import environment
2  import numpy as np
3  from parameters import *
4  import tensorflow as tf
5  from keras.layers import Dense, Input, Lambda, Add
6  from keras.models import Model
7  from keras import backend as K
```

Then, we initialize the "deep_q_learning_agent" class as follows:

---

deep_q_learning_agent.py

```
1  class deep_q_learning_agent:
2      def __init__(self, dueling):
3          self.env = environment()
```

```
4    self.action_history = []
5    self.state_history = []
6    self.reward_history = []
7    self.next_state_history = []
8
9
10   self.model = self.create_model()
11   self.target_model = self.create_model()
12
13   self.epsilon = 1.0
14   self.epsilon_min = 0.01
15   self.epsilon_decay = 0.9999
16   self.loss_function = tf.keras.losses.Huber()
17   self.optimizer = tf.keras.optimizers.Adam(
         learning_rate=learning_rate_deepQ)
```

In this "＿init＿" function, we first create a new environment and a
memory pool to store learning experiences, i.e. state, action, immediate reward,
and next state. The epsilon will be reduced from 1 to 0.01 with a decay factor
of 0.9999. The Q-network and the target Q-network are created by using the
"create_model" function below.

deep_q_learning_agent.py

```
1   def create_model(self, dueling):
2     input_shape = (num_features,)
3     X_input = Input(input_shape)   # input layer
4     X = X_input
5
6     X = Dense(512, input_shape=input_shape, activation="
         tanh")(X)   # first hidden layer
7
8     X = Dense(256, activation="tanh")(X)   # second
         hidden layer
9
10    X = Dense(64, activation="tanh")(X)   # third hidden
         layer
11
12    X = Dense(num_actions, activation="linear")(X)   #
         output layer
13
14    model = Model(inputs=X_input, outputs=X)   # create
         deep neural network
15    return model
```

In the "create_model" function, we first create an input layer with a shape
of "num_features"(which is the number of dimensions in the state space). The
hidden layers are then created by using the "Dense" class from Keras with the size,

the input shape, and the activation function. It is worth noting that using more hidden layers may improve the learning process. However, this also can increase the complexity of the deep neural network, resulting in long learning time. In addition, the overestimation problem often occurs when there are many hidden layers in the deep neural network.

Next, we provide the codes for storing experiences in the memory pool in the "remember" function below. In particular, the experience (state, action, immediate reward, next state) will be appended to four arrays, respectively. If the memory pool is full, the oldest experience will be removed.

---

deep_q_learning_agent.py

```
1  def remember(self, state, action, reward, next_state):
2      self.state_history.append(state)
3      self.action_history.append(action)
4      self.next_state_history.append(next_state)
5      self.reward_history.append(reward)
6      if len(self.reward_history) > memory_size:
7          del self.state_history[:1]
8          del self.action_history[:1]
9          del self.next_state_history[:1]
10         del self.reward_history[:1]
```

During the learning process, the deep Q-learning agent will randomly take some experiences in the memory pool and learn to update the deep neural network by using the "replay" function below.

---

deep_q_learning_agent.py

```
1   def replay(self):
2       indices = np.random.choice(range(len(self.
            state_history)), size= batch_size)
3       state_sample = np.array([self.state_history[i] for i
            in indices]).reshape((batch_size, num_features))
4       action_sample = np.array([self.action_history[i] for
            i in indices])
5       reward_sample = np.array([self.reward_history[i] for
            i in indices])
6       next_state_sample = np.array([self.
            next_state_history[i] for i in indices]).reshape
            ((batch_size, num_features))
7       future_rewards = self.target_model.predict
            (next_state_sample, verbose=0)
8       updated_q_values = reward_sample +
            discount_factor_deepQ * tf.reduce_max
            (future_rewards, axis=1)
9       mask = tf.one_hot(action_sample, num_actions)
10      with tf.GradientTape() as tape:
11          q_values = self.model(state_sample, training=False)
```

```
12    q_action = tf.reduce_sum(tf.multiply(q_values,
          mask), axis=1)
13    loss = self.loss_function(updated_q_values,
          q_action)
14
15  grads = tape.gradient(loss, self.model.
        trainable_variables)
16  self.optimizer.apply_gradients(zip(grads, self.model
        .trainable_variables))
```

In particular, we first generate random indices with the size of "batch_size."
Then, based on these indices, the state, action, next state, and reward samples
will be taken from the memory pool. The future reward is then obtained by feed-
ing the next state to the target Q-network. After that, the estimated Q-value is
calculated to construct the loss object. Finally, the gradient step is performed to
minimize the loss by using the optimizer defined in the "init" function. During
the learning process, the target Q-network is slowly updated with the weights of
the Q-network. This is done by using the following function:

deep_q_learning_agent.py

```
1  def target_update(self):
2    self.target_model.set_weights(self.model.get_weights())
```

In each learning step, given a system state, the deep Q-learning agent selects an
action based on the trained Q-network by using the following function:

deep_q_learning_agent.py

```
1  def get_action(self, state):
2    self.epsilon *= self.epsilon_decay
3    self.epsilon = max(self.epsilon_min, self.epsilon)
4    if np.random.random() < self.epsilon:
5      return np.random.choice(self.env.
            get_possible_action())
6    else:
7      list_value = self.model.predict(state, verbose=0)[0]
8      list_actions = self.env.get_possible_action()
9      max_q = -float("inf")
10     action = 0
11     for action_t in list_actions:
12       if list_value[action_t] >= max_q:
13         max_q = list_value[action_t]
14         action = action_t
15     return action
```

Similar to the Q-learning algorithm, the deep Q-learning agent also selects an action at each state based on the epsilon-greedy method. However, instead of using the Q-table, the deep Q-learning agent uses the deep Q-network to determine the action with the highest estimated Q-value.

Given the above functions, the deep Q-learning agent performs its learning process by using the "learning" function below:

```
deep_q_learning_agent.py
1   def learning(self):
2     total_reward = 0
3     for i in range(T):
4       current_state = self.env.get_state_deep()
5       current_state = np.reshape(current_state, (1,
            num_features))
6       action = self.get_action(current_state)
7
8       reward, next_state = self.env.perform_action_deep
            (action)
9       next_state = np.reshape(next_state, (1,
            num_features))
10      total_reward += reward
11      self.remember(current_state, action, reward,
            next_state)
12
13      self.replay()
14
15      if (i+1) % update_target_network == 0:
16        self.target_update()
17      if (i+1) % step == 0:
18        print("Iteration " + str(i + 1) + " reward: " +
              str(total_reward / (i + 1)))
```

Remark: It is worth noting that different problems can use the same structure of the deep Q-learning agent provided above. One just needs to change the environment class corresponding to a particular problem. The codes for the deep Q-learning agent can be kept the same with the modifications of the deep neural network and hyperparameters. In general, complex problems, e.g. those involving large state and action spaces, more random properties, and more conditions, will require more hidden layers in the deep neural network to better learn the system. In addition, other types of deep neural networks can also be used for special problems such as recurrent neural networks and convolutional neural networks.

4.3 Simulation Results and Performance Analysis

We will compare the performance of the deep Q-learning algorithm and the Q-learning algorithm using the parameters specified in the "parameters" function. First, we show the convergence rates of these algorithm in Figure 4.3.

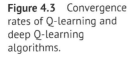

Figure 4.3 Convergence rates of Q-learning and deep Q-learning algorithms.

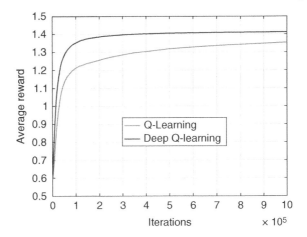

As can be observed, the convergence rate of the deep Q-learning algorithm is faster than that of the Q-learning algorithm. This is because the deep Q-learning algorithm, which uses a deep neural network, can learn experiences multiple times, allowing it to more efficiently understand the characteristics of the system.

It is worth noting that the performance of the deep Q-learning algorithm is greatly affected by the hyperparameters, e.g. learning rate, batch size, and deep neural network architecture. As such, it is essential to fine tune these parameters to obtain the best learning performance for each problem. In the following, we run the deep Q-learning algorithm with different parameters and observe its performance.

First, in Figure 4.4, we show the learning process of the deep Q-learning algorithm with different learning rates. Note that, in the deep Q-learning algorithm, the learning rate is the rate used by the optimizer (i.e. ADAM optimizer in our codes). As such, the learning rate has a strong impact on the learning process of the algorithm. As can be observed in Figure 4.4, the algorithm can quickly converge to the optimal policy with learning rates of 0.001 and 0.0001. When we increase the learning rate to 0.01 and 0.1, the algorithm cannot converge to the optimal policy, even after 10^6 learning iterations.

Next, in Figure 4.5, we show the performance of the deep Q-learning algorithm with different decay factors. The decay factor in the epsilon greedy method is a very important parameter in reinforcement learning since it has a strong impact on the exploration process. In particular, the reduction rate of the epsilon value depends on the decay factor. The higher the epsilon value is, the higher chances the reinforcement learning process can explore the system. However, if the epsilon value is too high, the reinforcement learning process may not be able to effectively use the exploitation process, resulting in a low convergence rate. As can be seen in

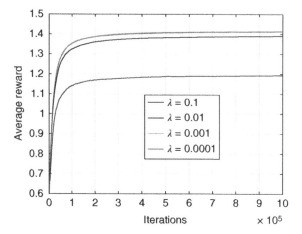

Figure 4.4 Convergence rates of deep Q-learning with different learning rates.

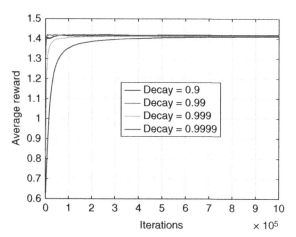

Figure 4.5 Convergence rates of deep Q-learning with different decay factors.

Figure 4.5, with the decay factor of 0.9999, the algorithm converges much slower compared to the cases with decay factors of 0.999 and 0.99.

In Figure 4.6, we show the learning process of the deep Q-learning algorithm with different activation functions. As can be observed in the figure, with *Relu* activation function [11], the learning process is not stable. It first converges to the optimal policy and then drops when the algorithm keeps learning. In contrast, *Tanh* activation function [11] give a more stable straining process. Therefore, it is essential to select an appropriate activation function for the deep neural

Figure 4.6 Convergence rates of deep Q-learning with different activation functions.

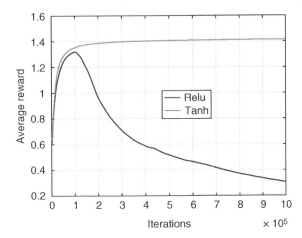

Figure 4.7 Convergence rates of deep Q-learning with different optimizers.

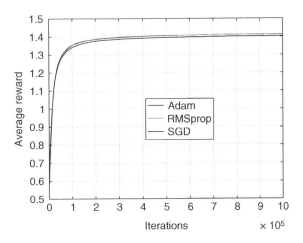

network. Different problems may require different activation functions. Based on our experience, *Tanh* activation function can be used in most reinforcement learning problems.

In Figure 4.7, we show the learning process of the DRL method with different optimizers, including Adam [12], RMSprop [13], and conventional stochastic gradient descent (SGD) optimizer [11]. As can be seen in the figure, while the Adam and RMSprop can quickly converge, the conventional SGD optimizer cannot converge to the optimal policy after 10^6 training steps. The reason is that Adam and RMSprop are advanced extensions of SGD that can learn with different learning rates in each iteration and weight. In DRL, Adam is the most popular and useful optimizer.

4.4 Chapter Summary

In this chapter, we have presented the detailed implementation of Q-learning and deep Q-learning models to address the anti-jamming problem in wireless communications. In particular, we have considered a wireless system in which a jammer aims at disrupting the legitimate communications between a transmitter and a receiver. As the jamming strategies are difficult to obtain in advance, dealing with jamming attacks is very challenging. DRL is an effective solution to deal with jamming attacks. As such, we have first formulated the anti-jamming problem as an MDP. Then, we have developed the detailed codes for the Q-learning algorithm. After that, the codes of DRL have been provided with detailed explanations. Finally, the performances of the DRL methods have been discussed, under different hyperparameter settings.

References

1 N. Van Huynh, D. N. Nguyen, D. T. Hoang, and E. Dutkiewicz, ""Jam me if you can": Defeating jammer with deep dueling neural network architecture and ambient backscattering augmented communications," *IEEE Journal on Selected Areas in Communications*, vol. 37, no. 11, pp. 2603–2620, 2019.

2 N. Van Huynh, D. N. Nguyen, D. T. Hoang, E. Dutkiewicz, and M. Mueck, "Ambient backscatter: A novel method to defend jamming attacks for wireless networks," *IEEE Wireless Communications Letters*, vol. 9, no. 2, pp. 175–178, 2019.

3 N. Van Huynh, D. T. Hoang, D. N. Nguyen, and E. Dutkiewicz, "DeepFake: Deep dueling-based deception strategy to defeat reactive jammers," *IEEE Transactions on Wireless Communications*, vol. 20, no. 10, pp. 6898–6914, 2021.

4 M. K. Hanawal, M. J. Abdel-Rahman, and M. Krunz, "Joint adaptation of frequency hopping and transmission rate for anti-jamming wireless systems," *IEEE Transactions on Mobile Computing*, vol. 15, no. 9, pp. 2247–2259, 2015.

5 W. Xu, W. Trappe, Y. Zhang, and T. Wood, "The feasibility of launching and detecting jamming attacks in wireless networks," in *Proceedings of the 6th ACM International Symposium on Mobile ad hoc Networking and Computing*, pp. 46–57, 2005.

6 A. Mpitziopoulos, D. Gavalas, C. Konstantopoulos, and G. Pantziou, "A survey on jamming attacks and countermeasures in WSNs," *IEEE Communications Surveys & Tutorials*, vol. 11, no. 4, pp. 42–56, 2009.

7 L. Jia, Y. Xu, Y. Sun, S. Feng, and A. Anpalagan, "Stackelberg game approaches for anti-jamming defence in wireless networks," *IEEE Wireless Communications*, vol. 25, no. 6, pp. 120–128, 2018.

8 V. Liu, A. Parks, V. Talla, S. Gollakota, D. Wetherall, and J. R. Smith, "Ambient backscatter: Wireless communication out of thin air," *ACM SIGCOMM Computer Communication Review*, vol. 43, no. 4, pp. 39–50, 2013.

9 K. Firouzbakht, G. Noubir, and M. Salehi, "On the capacity of rate-adaptive packetized wireless communication links under jamming," in *Proceedings of the 5th ACM Conference on Security and Privacy in Wireless and Mobile Networks*, pp. 3–14, 2012.

10 J. Filar and K. Vrieze, *Competitive Markov decision processes*. Springer Science & Business Media, 2012.

11 I. Goodfellow, Y. Bengio, and A. Courville, *Deep learning*. MIT Press, 2016.

12 D. P. Kingma and J. Ba, "Adam: A method for stochastic optimization," *arXiv preprint arXiv:1412.6980*, 2014.

13 G. Hinton, N. Srivastava, and K. Swersky, "Neural networks for machine learning lecture 6a overview of mini-batch gradient descent," [Online]. Available at: http://www.cs.toronto.edu/~hinton/coursera/lecture6/lec6.pdf.

Part II

Applications of DRL in Wireless Communications and Networking

5

DRL at the Physical Layer

5.1 Beamforming, Signal Detection, and Decoding

In this section, we present a comprehensive review on the applications of deep reinforcement learning (DRL) in beamforming, signal detection, and channel decoding.

5.1.1 Beamforming

5.1.1.1 Beamforming Optimization Problem

Beamforming is a radio frequency (RF) transmission technology that directs a wireless signal toward a specific device, instead of broadcasting it to all directions. By doing this, beamforming can help to improve the robustness and data rate of the connection. Moreover, it can help to mitigate interference between devices as signals are only focused where needed. Nevertheless, beamforming requires more computing resources and energy for calculations of the beamforming coefficients or vectors (e.g. for systems with multiple antennas). In addition, the problem of beamforming optimization is usually nonconvex.

In the following, we present a standard beamforming optimization problem [1]. Consider a cellular network with L base stations (BSs) where each BS is equipped with a uniform linear array of M antennas. Each user equipment (UE) is assumed to have a single antenna and is served by only one BS. Considering interference from nearby BSs, the received signals from the lth BS at the UE can be expressed as follows:

$$y_l = \mathbf{h}_{l,l}^* \mathbf{f}_l x_l + \sum_{b \neq l} \mathbf{h}_{l,b}^* \mathbf{f}_b x_b + n_l, \tag{5.1}$$

where $x_l \in \mathbb{C}$ and $x_b \in \mathbb{C}$ are the transmitted signals from the lth BS and the bth BS, respectively. $\mathbf{f}_l, \mathbf{f}_b \in \mathbb{C}^{M \times 1}$ present the downlink beamforming vector at the lth BS and the bth BS, respectively. Each beamforming vector is selected from a

Deep Reinforcement Learning for Wireless Communications and Networking:
Theory, Applications, and Implementation, First Edition.
Dinh Thai Hoang, Nguyen Van Huynh, Diep N. Nguyen, Ekram Hossain, and Dusit Niyato.
© 2023 The Institute of Electrical and Electronics Engineers, Inc. Published 2023 by John Wiley & Sons, Inc.

beamsteering-based beamforming codebook $\mathcal{F}$ with the nth element defined as follows:

$$\mathbf{f}_n := \mathbf{a}(\theta_n) = \frac{1}{\sqrt{M}}\left[1, e^{jkd\cos(\theta_n)}, \ldots, e^{jkd(M-1)\cos(\theta_n)}\right]^\top, \tag{5.2}$$

where θ_n is the steering angle, and d and k present the antenna spacing and the wave number, respectively. $\mathbf{a}_n$ represents the array steering vector with direction θ_n which is calculated by dividing the antenna angular space by the number of antennas M.

Here, $\mathbf{h}_{l,l}, \mathbf{h}_{l,b} \in \mathbb{C}^{M\times1}$ denote the channel vectors that connect the UE to the lth and bth BSs, respectively. Also, $n_l \sim \text{Normal}(0, \sigma_n^2)$ is the noise following a complex Normal distribution with zero mean and variance σ_n^2. Assuming that millimeter wave (mmWave) technology is adopted, the downlink channel from BS b to the user in BS l can be expressed as

$$\mathbf{h}_{l,b} = \frac{\sqrt{M}}{\rho_{l,b}} \sum_p^{N_{l,b}^p} \alpha_{l,b}^p \mathbf{a}^*(\theta_{l,b}^p), \tag{5.3}$$

where $\rho_{l,b}$ is the path-loss between BS b and the user served in the area of BS l. $\alpha_{l,b}^p$ and $\theta_{l,b}^p$ are the complex path gain and the angle of departure (AoD) of the pth path between BS b and the user, respectively. $\mathbf{a}(\theta_{l,b}^p)$ represents the array response vector corresponding to the AoD $\theta_{l,b}^p$. $N_{l,b}^p$ denotes the number of channel paths between BS b and the user.

Given the above, we can express the received downlink power measured by the UE at a given time t from BS b, i.e. $P_{\text{UE}}^{l,b}[t]$ and BS l, i.e. $P_{\text{UE}}^{l,l}[t]$ as follows:

$$\begin{aligned} P_{\text{UE}}^{l,b}[t] &= P_{\text{TX},b}[t]\left|\mathbf{h}_{l,b}^*[t]\mathbf{f}_b[t]\right|^2, \\ P_{\text{UE}}^{l,l}[t] &= P_{\text{TX},l}[t]\left|\mathbf{h}_{l,l}^*[t]\mathbf{f}_l[t]\right|^2, \end{aligned} \tag{5.4}$$

where $P_{\text{TX},b}$ and $P_{\text{TX},l}$ are the transmit power of BS b and BS l, respectively. We then can formulate the received signal-to-interference-plus-noise ratio (SINR) of the UE served by BS l at time step t as follows:

$$\begin{aligned} \gamma^l[t] &= \frac{P_{\text{UE}}^{l,l}[t]}{\sigma_n^2 + \sum_{b\neq l} P_{\text{UE}}^{l,b}[t]} \\ &= \frac{P_{\text{TX},l}[t]|\mathbf{h}_{l,l}^*[t]\mathbf{f}_l[t]|^2}{\sigma_n^2 + \sum_{b\neq l} P_{\text{TX},b}[t]|\mathbf{h}_{l,b}^*[t]\mathbf{f}_b[t]|^2}. \end{aligned} \tag{5.5}$$

To maximize the SINR at the UE, we need to solve the following optimization problem:

$$\underset{\substack{P_{\text{TX},j}, \forall j \\ \mathbf{f}_j[t], \forall j}}{\text{maximize}} \sum_{j\in\{1,2,\ldots,L\}} \gamma^j[t], \tag{5.6a}$$

subject to $P_{\text{TX},j}[t] \in \mathcal{P}, \quad \forall j$ (5.6b)

$$\mathbf{f}_j[t] \in \mathcal{F}, \quad \forall j \tag{5.6c}$$

$$\gamma^j[t] \geq \gamma_{\text{target}}, \tag{5.6d}$$

where γ_{target} is the target SINR of the downlink. Note that, the above optimization problem also aims at optimizing the transmit power at the BSs together with beamforming. As such, we have the first constraint (i.e. (5.6b)), in which $\mathcal{P}$ is the set of potential transmit power levels.

As observed, the optimization problem is nonconvex due to the nonconvexity of constraints (5.6b) and (5.6c) [1]. Consequently, it is difficult if not impossible to obtain the optimal beamforming solution with conventional approaches (e.g. exhaustive search). DRL is a useful tool to address this challenge and obtain efficient suboptimal, solutions. Over the last few years, DRL has been widely and successfully adopted to solve the beamforming problem in different types of wireless systems [2–12], as discussed in the following.

5.1.1.2 DRL-Based Beamforming
In [2], the authors considered the beamforming problem to mitigate intercell interference for downlink transmissions in multicell multi-input single-output (MISO) systems. The traditional approaches are generally impractical due to the high computational complexity and overhead for acquiring channel state information (CSI). Therefore, the authors proposed to deploy a DRL agent at each BS to observe the environment information (i.e. state), including its local information, interferer information, and its neighbors' information, to make a beamforming action. Through the reinforcement learning (RL) process, BSs can learn and perform optimal beamforming actions given particular system states. The simulation results demonstrated that the proposed DRL-based beamforming solution can achieve a similar system capacity as the ideal fractional programming algorithm while requiring much lower system overhead. Similarly, the authors in [5] also demonstrated the power of DRL in addressing the beamforming problem for multiuser MISO systems. However, conventional DRL approaches cannot work in the case where there are many active users in the same cell or large-scale antennas are deployed at the BS. This is because the discrete action space expands exponentially with the number of antennas and users. Consequently, the size of the output layer will be very large, resulting in poor training performance. For this reason, the authors in [7] introduced a multi-agent deep deterministic policy gradient (DDPG) algorithm, an extended version of DRL, to handle the continuous action space and obtain better beamforming performance compared to other solutions. The DDPG algorithm was adopted in [8] to optimize

beamforming at the uplink of cell-free networks. In cell-free networks, a large number of mobile devices are served by multiple BSs simultaneously using the same time/frequency resources, and thus creating high-signal processing demands at receivers and transmitters. To address this problem, the authors proposed a centralized DDPG algorithm and then enhanced this method by leveraging experiences collected from different access points (APs) based on the distributed distributional deterministic policy gradients algorithm (D4PG). In the proposed distributed algorithm, the APs act as agents in the DRL model, and thus the beamforming computations can be divided among distributed agents, i.e. APs.

Recently, DRL has also been widely adopted for reconfigurable intelligent surface (RIS)-aided wireless communications systems to improve the communication efficiency. An RIS is equipped with a large array of passive scattering elements controlled by an RIS controller as illustrated in Figure 5.1. By optimizing the reflecting coefficients of all scattering elements (i.e. passive beamforming), an RIS can enhance the signal strength at the receiver. In [10], the authors jointly optimized the active beamforming strategy from an AP and the passive beamforming of the RIS. In particular, the RIS is equipped with N reflecting elements assisted information transmissions from the M-antenna AP to the single-antenna receiver as shown in Figure 5.1. The RIS's passive beamforming is denoted as $\boldsymbol{\Theta} = \mathrm{diag}(\rho_1 e^{j\theta_1}, \ldots, \rho_n e^{j\theta_n}, \ldots, \rho_N e^{j\theta_N})$, in which $\theta_n \in [0, 2\pi]$ is the phase shift and $\rho_n \in [0,1]$ is the magnitude. Then, the RIS-assisted equivalent channel from the AP to the receiver can be formulated as $\hat{\mathbf{g}} = \mathbf{g} + \mathbf{H}\boldsymbol{\Theta}\mathbf{f}$, where $\mathbf{g} \in \mathbb{C}^{M \times 1}$, $\mathbf{H} \in \mathbb{C}^{M \times N}$, and $\mathbf{f} \in \mathbb{C}^{N \times 1}$ are the complex channel from the AP to the

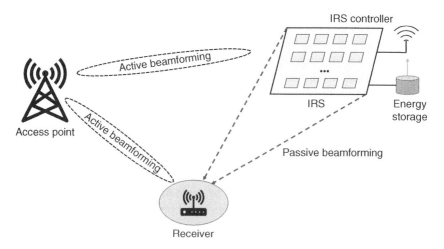

Figure 5.1 RIS-assisted MISO systems. Adapted from Lin et al. [10].

receiver, the AP to the RIS, and the RIS to the receiver, respectively. Given the above, we can express the received signals at the receiver as follows:

$$y = \hat{\mathbf{g}}^{\mathrm{H}}\mathbf{w}s + v, \tag{5.7}$$

where $\hat{\mathbf{g}}^{\mathrm{H}}$ is the conjugate transpose of $\hat{\mathbf{g}}$, $\mathbf{w}$ is the AP's beamforming vector, s is the complex symbol with unit transmit power, and v is the Gaussian noise with zero mean and normalized variance. Then, the received signal-to-noise ratio (SNR) at the receiver can be expressed as

$$\gamma(\mathbf{w}, \boldsymbol{\Theta}) = \|(\mathbf{g} + \mathbf{H}\boldsymbol{\Theta}\mathbf{f})^{\mathrm{H}}\mathbf{w}\|^2. \tag{5.8}$$

Considering a practical case in which all the reflecting elements have the same power-splitting ratio ρ, the received SNR can be reformulated as:

$$\gamma(\mathbf{w}, \boldsymbol{\theta}) = \|(\mathbf{g} + \rho\mathbf{H}_{\mathbf{f}}\boldsymbol{\theta})^{\mathrm{H}}\mathbf{w}\|^2, \tag{5.9}$$

where $\boldsymbol{\theta} = [e^{j\theta_1}, \dots, e^{j\theta_N}]^{\mathrm{T}}$ is the phase vector, $\mathbf{H}_{\mathbf{f}}$ is the cascaded AP-RIS-receiver channel defined as $\mathbf{H}_{\mathbf{f}} \triangleq \mathrm{diag}(\mathbf{f})\mathbf{H}$. The authors aimed at minimizing the AP's transmit power $\|\mathbf{w}\|^2$ by jointly optimizing the active and passive beamforming strategies with the constraints of RIS's power and target SNR as follows:

$$\underset{\mathbf{w}, \boldsymbol{\theta}, \rho}{\text{minimize}} \ \|\mathbf{w}\|^2, \tag{5.10a}$$

$$\text{subject to } \|(\mathbf{g} + \rho\mathbf{H}_{\mathbf{f}}\boldsymbol{\theta})^{\mathrm{H}}\mathbf{w}\|^2 \geq \gamma^{\dagger}, \quad \forall \mathbf{H}_{\mathbf{f}} \in \mathbb{U}_{\mathbf{f}} \tag{5.10b}$$

$$\eta(1 - \rho^2)\|\mathbf{H}^{\mathrm{H}}\mathbf{w}\|^2 \geq N\mu, \quad \forall \mathbf{H} \in \mathbb{U}_{\mathbf{h}} \tag{5.10c}$$

$$\rho \in (0,1) \text{ and } \theta_n \in (0, 2\pi), \quad \forall n \in \mathcal{N}, \tag{5.10d}$$

where $\mathbb{U}_{\mathbf{f}}$ and $\mathbb{U}_{\mathbf{h}}$ are the uncertainty sets of for the AP-RIS and AP-RIS-receiver channels. η is the power harvesting efficiency and μ is the power consumption of a single scattering element. Clearly, the above joint optimization problem is nonconvex as the power splitting ratio is coupled with the phase vector. Moreover, constraints (5.10c) and (5.10d) are semi-infinite constraints. As such, it is very difficult to solve the optimization problem efficiently with conventional approaches. For that, the authors proposed a DRL-based solution to obtain the optimal beamforming strategies for both the AP and the RIS.

Interestingly, to improve the performance of the proposed DRL-based solution, the authors integrated a model-based optimization into the model-free DRL algorithm. To do that, when the DRL algorithm generates a part of the action (i.e. passive beamforming), the optimization module will find the other part of the action (i.e. active beamforming) by solving an approximate convex problem. The simulation results demonstrated that the proposed solution can not only speed up the learning process but also can significantly improve the system performance compared to conventional methods.

The authors in [11] also considered the problem of joint passive and active beamforming to secure wireless communication systems. Specifically, a DRL-based beamforming solution is proposed to obtain the joint beamforming policy that can maximize the received SINR at the legitimate users and suppress the wiretapped rate at the eavesdroppers. Moreover, different quality-of-service requirements and time-varying channel conditions are also considered in the learning process of the proposed solution. The simulation results demonstrate that by using DRL, a much higher system secrecy rate can be achieved compared to the existing approaches. Similarly, the authors in [12] also proposed to use DRL to jointly optimize the active and passive beamforming policies as well as the UAV's trajectory to maximize the average secrecy rate of the system under the presence of eavesdroppers.

mmWave communication is a key technology for high throughput and low latency thanks to the use of frequency bands above 24 GHz. Nevertheless, transmitting data over these frequencies makes mmWave technology vulnerable to path blocking as well as mobility. Several beamforming techniques have been proposed to address these problems. Unfortunately, conventional optimization approaches cannot effectively deal with large-scale antenna arrays and multiple users under dynamic channel environment and user mobility. DRL is a promising solution to address this issue [13–19]. In [14], the authors proposed PrecoderNet, a novel DRL-based method, to optimize the beamforming process for mmWave systems. In particular, the authors considered a hybrid beamforming strategy in which the fully digital beamformer is replaced by a low-dimensional digital precoder and a high-dimensional digital precoder. Then, a DRL algorithm is proposed to obtain the optimal hybrid beamforming policy. The DRL agent considers an action as the digital beamformer and analog combiner, the current state is the action in the previous training step, and the reward is the spectral efficiency. The simulation results then demonstrated the effectiveness of PrecoderNet in spectral efficiency, time consumption, and bit error rate (BER). Moreover, it is also robust to imperfect CSI, making it feasible to implement in many real-world settings. Hybrid beamforming based on DRL was also studied in [15] and [16] for multiuser mmWave wireless systems. More importantly, the authors in [16] introduced a multi-agent DRL method in which the agents jointly explore the environment, resulting in a much higher convergence rate, compared to conventional approaches.

The authors in [17] proposed a DRL-based framework to quickly obtain the optimal beamforming policy in high mobility mmWave systems. Specifically, the authors considered a mmWave system with multiple APs and several fast-moving users (e.g. cars). In this system, each user can be served by several APs

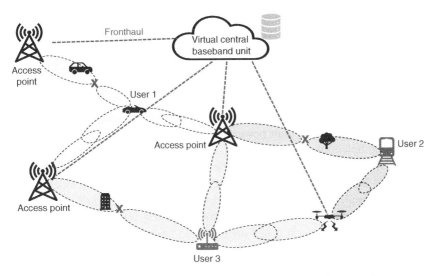

Figure 5.2 Beamforming for high mobility mmWave systems. Fozi et al. [17].

simultaneously as shown in Figure 5.2. As mmWave transmissions are extremely vulnerable to mobility, obtaining the optimal beamforming to guarantee continuous services for all users is very challenging. In addition, fast beamforming is also vital when users move to new zones. To address this problem, the authors utilized DRL to develop a fast and efficient beamforming framework. In particular, the system is first modeled as a partially observed Markov decision process (MDP) in which the state is defined by the instantaneous CSI, the action set consists of precoding and beamforming matrices, and the reward function is defined based on the quality of service (QoS) metric. Both centralized and distributed DRL methods are studied. The simulation results demonstrated that the proposed mmWave beamforming framework can achieve better performance than existing schemes while requiring minimal overhead and complexity. In [18], the authors considered a more challenging scenario in which the transmitters, namely, the remote radio heads, aim to perform beamforming to provide reliable communication for high-speed trains in a rail network. Due to the very high speed, the handover period will be considerably short. Moreover, the rapid relative movement between trains can cause serious Doppler shifts and small channel coherence time. As a result, obtaining the optimal beamforming strategy is vital yet very challenging for communications in rail networks. Motivated by this, the authors in [18] proposed an effective beamforming scheme based on DRL to train and determine the optimal beam direction for trains to maximize their received signal power. The DRL agent considers the system state as the position of the train receiver, the action space is the optimal beam direction, and the reward function is the received signal power. The simulation results demonstrated that the proposed solution can outperform several schemes in the literature.

Recently, several works have focused on the joint beamforming and power control problem [1, 20–25]. Power control is a technique that allows the BSs to adjust their transmission power to mitigate interference. By jointly optimizing the power control and beamforming policies, the system performance can be significantly enhanced. For example, in [20], the authors aimed at maximizing the sum-rate of users by a joint power control and beamforming method for the time-varying downlink channel. A DRL-based approach was developed to deal with the dynamics and uncertainty of the system and obtain the optimal power control and beamforming policy without requiring CSI. Similarly, the authors in [22] also used DRL to jointly optimize the power control and beamforming strategies for Internet-of-Things (IoT) networks. In [23], a joint power allocation and beamforming approach based on DRL was proposed to optimize the total user sum-rate in mmWave downlink nonorthogonal multiple access (NOMA) systems.

In [1], the authors further improved the performance of joint power control and beamforming approaches by also taking into account interference coordination. The inter-cell interference and inter-beam interference significantly impact the system performance. However, designing a method that can (i) jointly solve the power control, interference coordination, and beamforming problems; (ii) achieve the desired SINR; and (iii) minimize the size of the action space is very challenging. To address this challenge, the authors in [1] proposed a DRL-based solution to explore the solution space by learning from interactions with the environment. To reduce the number of actions that need to be learned by the agent, the authors proposed a binary encoding technique to encode relevant actions into one action. The simulation results demonstrated that the proposed solution can outperform conventional reinforcement learning approaches as well as the industry standard power allocation scheme. Nevertheless, the authors aimed to reduce the amount of feedback from users, resulting in a low convergence rate of the proposed DRL algorithm [26]. To overcome this problem, advanced DRL models, e.g. deep dueling, can be adopted.

5.1.2 Signal Detection and Channel Estimation

5.1.2.1 Signal Detection and Channel Estimation Problem

Besides beamforming, DRL can also be applied to enhance the channel estimation and signal detection process in wireless communications systems. Fundamentally, channel estimation is the process of estimating the CSI, while signal detection is the process of estimating the transmitted signals. The signal detection process can be significantly enhanced with the support of the CSI. Many works have addressed the problem of optimizing the performance of channel estimation and signal detection processes. Nevertheless, conventional approaches require complex mathematical models to express the channel properties as well as construct detection hypotheses. In addition, they usually require prior information that may not be available in advance in practice.

In the following, we present a standard channel estimation and signal detection problem in wireless communications. Specifically, the received signal at time slot n of a wireless communication system can be expressed as follows [27, 28]:

$$\mathbf{y}[n] = \mathbf{H}\mathbf{x}[n] + \mathbf{z}[n], \tag{5.11}$$

where $\mathbf{H} \in \mathbb{C}^{N_{rx} \times N_{tx}}$ is the channel between the transmitter and the receiver, N_{tx} and N_{rx} are the number of transmit and receive antennas, respectively. $\mathbf{x}[n] \in \mathcal{X}^{N_{tx}}$ denotes the data symbol vector transmitted at time n, and $\mathbf{z}[n]$ is a circularly symmetric complex Gaussian noise vector at time n, i.e. $\mathbf{z}[n] \sim \mathcal{CN}(\mathbf{0}_{N_{rx}}, \sigma^2 \mathbf{I}_{N_{rx}})$.

Denote $\mathbf{x}_k$ is a vector in $\mathcal{X}^{N_{tx}}$ with $k \in \mathcal{K} = \{1, \ldots, K\}$, where $K = |\mathcal{X}^{N_{tx}}|$, the a-posteriori-probability (APP) of event $\{\mathbf{x}[n] = \mathbf{x}_k\}$ given received signal $\mathbf{y}[n]$ can be calculated as follows [27, 28]:

$$\theta_k[n] = \frac{\exp\left(-\frac{1}{\sigma^2}\|\mathbf{y}[n] - \mathbf{H}\mathbf{x}_k\|^2\right)}{\sum_{j \in \mathcal{K}} \exp\left(-\frac{1}{\sigma^2}\|\mathbf{y}[n] - \mathbf{H}\mathbf{x}_j\|^2\right)}. \tag{5.12}$$

In practice, it is impossible to obtain the exact APP in (5.12) as the perfect channel information is not available to the receiver. To address this problem, the channel information is usually estimated by using pilot signals. Similar to (5.11), the received pilot signals at the receiver can be expressed as follows:

$$\mathbf{y}_p[n] = \mathbf{H}\mathbf{p}[n] + \mathbf{z}[n], \tag{5.13}$$

where $\mathbf{p}[n]$ is the pilot signal (which is known by the receiver) sent at time slot n. Then, the concatenated received pilot signals during the pilot transmission can be obtained as follows:

$$\mathbf{Y}_p = \mathbf{H}\mathbf{P} + \mathbf{Z}_p, \tag{5.14}$$

where $\mathbf{P} = [\mathbf{p}[-T_p + 1], \ldots, \mathbf{p}_0]$ and $\mathbf{Z}_p = [\mathbf{z}[-T_p + 1], \ldots, \mathbf{z}_0]$, with T_p is the length of the pilot block. In practice, there are several methods to estimate the channel based on pilot signals, of which the linear minimum-mean-squared-error (LMMSE) is widely used in the literature. In particular, based on (5.14), the LMMSE filter $\mathbf{W}_{LMMSE}$ of the system can be expressed as [27, 28]:

$$\mathbf{W}_{LMMSE} = \mathbf{P}^H (\mathbf{P}\mathbf{P}^H + \sigma^2 \mathbf{I}_{N_{tx}})^{-1}. \tag{5.15}$$

Then, the LMMSE channel estimate can be calculated as follows:

$$\hat{\mathbf{H}}_p = \mathbf{Y}_p \mathbf{W}_{LMMSE} = \mathbf{Y}_p \mathbf{P}^H (\mathbf{P}\mathbf{P}^H + \sigma^2 \mathbf{I}_{N_{tx}})^{-1}. \tag{5.16}$$

Based on (5.16), the approximate APP can be expressed as follows [27, 28]:

$$\hat{\theta}_k[n] = \frac{\exp\left(-\frac{1}{\sigma^2}\|\mathbf{y}[n] - \hat{\mathbf{H}}_p\mathbf{x}_k\|^2\right)}{\sum_{j\in\mathcal{K}}\exp\left(-\frac{1}{\sigma^2}\|\mathbf{y}[n] - \hat{\mathbf{H}}_p\mathbf{x}_j\|^2\right)}, \tag{5.17}$$

where $\mathbf{x}_k$ is a vector in $\mathcal{X}^{N_{tx}}$ with $k \in \mathcal{K} = \{1, \dots, K\}$ where $K = |\mathcal{X}^{N_{tx}}|$. From (5.17), the transmitted signal can be detected. Nevertheless, the performance of the above channel estimation and signal detection greatly depends on the number of pilot signals. With insufficient number of pilot signals, there will be channel estimation errors (i.e. $\hat{\mathbf{H}}_p - \mathbf{H}$) at the receiver, and thus reducing its detection performance [27, 28]. Unfortunately, in practice, the number of pilot signals should not be too large as it will significantly reduce the communication efficiency, i.e. less information is transmitted. To address this problem, reinforcement learning can be used to aid the channel estimation and signal detection process [27, 29–35], as discussed in Section 5.1.2.2.

5.1.2.2 RL-Based Approaches

To support the channel estimation and signal detection with limited number of pilot signals, the authors in [27] proposed an idea of using detected symbol vectors as additional pilot signals for updating the channel estimate in (5.16). However, due to the dynamics and uncertainty of the wireless environment, there will be detected symbol vectors that are not good candidates for pilot signals. As such, the authors developed an RL algorithm to help the receiver learn and obtain the optimal set of detected symbol vectors that can minimize the channel estimation error. To do that, the authors first formulated the problem as an MDP in which the state consists of pilot signals and the detected symbol vectors, the action set consists of binary actions to decide whether the current detected symbol vector is selected as additional pilot signal or not, and the reward is the mean-square-error reduction. Then, an RL algorithm is developed to efficiently obtain the optimal policy when selecting detected symbol vectors based on Monte Carlo tree search method. The simulation results revealed that the proposed solution can effectively reduce the channel estimation error and the loss in detection performance.

Similarly, in [31], the authors proposed a signal detection method based on reinforcement learning for multi-input multi-output (MIMO) systems with one-bit analog-to-digital converters (ADCs) as shown in Figure 5.3. In particular, one-bit ADCs are introduced as a promising technique to reduce the power consumption caused by high-precision ADCs. However, in MIMO systems, one-bit ADCs need the knowledge of likelihood functions to obtain high detection accuracy. Due to the coarse quantization at the ADCs, there will be a mismatch in the likelihood function at the receiver. As such, the authors proposed a likelihood function estimation method based on reinforcement learning that is robust to

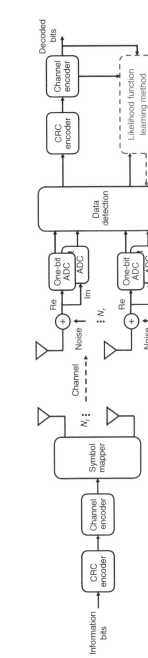

Figure 5.3 An MIMO system with one-bit ADCs. Adapted from Jeon et al. [31].

the mismatch caused by the use of ADCs. As shown in Figure 5.3, the proposed method for learning the likelihood function is used to improve the accuracy of the signal detection by learning from input–output samples (states) sent from the data detection block and the decoder block. By learning this information, the agent decides whether the current input–output sample is exploited to update the empirical likelihood function or not. In this way, the accuracy of the likelihood function can be significantly improved as demonstrated in the simulation results. Similarly, the authors in [33] also aimed at improving the signal detection accuracy in massive MIMO systems with one-bit ADCs by using reinforcement learning.

The authors in [29] developed a frequency-domain denoising method based on reinforcement learning for channel estimation in MIMO orthogonal frequency-division multiplexing (OFDM) systems without requiring prior CSI and labeled data. Specifically, the authors considered the denoising process as the problem of determining the optimal sequential order on OFDM subcarriers to reduce the mean square error of the estimation method. The denoising process is then formulated as an MDP in which the system state is defined as a set of channel estimates, the action is defined as the index of the channel estimate that will be denoised, and the reward function is defined as the noise reduction. The simulation results demonstrated that the proposed solution can achieve higher estimation performance compared to other schemes in the literature. Similarly, in [34], the authors proposed a DRL-based approach for end-to-end multiuser channel estimation and beamforming. By considering the received pilot signals, the history of channel prediction, the pilot sequence, and the feedback channel information as the system state, the agent can learn vital properties of the system to obtain the optimal policy for real-time CSI estimation and beamforming. In this way, the system performance in terms of sum-rate can be significantly improved. It is worth noting that the proposed solution can work well without the prior assumption of perfect CSI or large training CSI data set.

5.1.3 Channel Decoding

Besides channel estimation and signal detection, reinforcement learning can also be applied for channel encoding and decoding in wireless communications [36–43]. In [36], the authors proposed to use reinforcement learning to determine the best decoding strategies for binary linear codes. To formulate the decoding problem as an MDP, the bit-flipping (BF) decoding technique was adopted. In particular, this technique can construct a suitable metric that allows the decoder to rank the bits based on their reliability under code constraints. Then, the reinforcement-learning algorithm is proposed to select the best bit to be flipped in the received word at each time slot. The simulation results demonstrated the effectiveness of the proposed solution. Similarly, the authors in [37] developed a novel bit-flipping algorithm to decode polar codes. The parameters of the proposed bit-flipping algorithm were then optimized to improve the error-correction performance. To do that, the parameter optimization problem was modeled as an on-policy reinforcement-learning problem and solved by using RL techniques.

In [38], the authors formulated the polar code construction problem as a maze-traversing game that can be solved by reinforcement-learning techniques. Different from existing polar code construction methods, the proposed solution can determine the input bits that should be frozen in a purely sequential manner, instead of sorting and selecting bit-channels by their reliability. The simulation results demonstrated that the proposed solution can obtain code construction policies that achieve better frame error rate performance compared to other state-of-the-art solutions. Decoding polar codes with reinforcement learning was also extensively studied in [39] and [40].

Reinforcement learning can also be used to enhance decoding of sparse graph-based channel codes. In [41], the authors demonstrated that reinforcement learning can be applied to decode short-to-moderate-length sparse graph-based channel codes. In particular, they considered low-density parity check (LDPC) codes which have been standardized for data communications in 5G. In practice, the LDPC codes can be decoded via belief propagation by constructing sparse bipartite graphs with two types of vertex sets: (i) check nodes and (ii) variable nodes. These nodes will be updated during the decoding process. The authors proposed to use reinforcement learning to obtain the optimal check node scheduling policy to improve the decoding performance. Specifically, the problem is first formulated as a multi-armed bandit problem and then solved by the Q-learning algorithm. The simulation results revealed that the proposed solution can significantly improve the error-correction performance. However, the training complexity of this solution is high due to the large number of states. To address this issue, the authors in [42] developed a new approach, named RELDEC, that uses a clustering technique and a meta Q-learning algorithm for dynamic scheduling. The simulation results revealed that RELDEC can outperform several decoding techniques for LDPC codes, including codes designed for the 5G New Radio (NR) standard.

5.2 Power and Rate Control

5.2.1 Power and Rate Control Problem

In wireless communications, mitigating radio interference plays a vital role since interference among devices can significantly degrade the system performance, and consequently, the QoS [44]. One potential approach to mitigate interference is power control. In particular, power control allows a transmitter to adjust its transmit power to mitigate the effect of interference it may cause to other wireless devices. In addition, different users may require different transmission rates to satisfy their service requirements [45], e.g. augmented reality (AR) and virtual reality (VR) services may require much higher transmission rates than nonreal time services. As such, optimizing transmission rates for a huge number of users is very challenging. Conventional approaches often adopt convex optimization solutions to obtain the optimal power and rate control policies.

As an example, let's consider a cellular system with C cells $\{1, \ldots, c, \ldots, C\}$. Each cell can serve a set of users denoted by $\mathcal{N}_c$. Then, the total number of users in the system can be calculated as $N = \sum_c N_c$, with $N_c = |\mathcal{N}_c|$. Denote P_c as the downlink transmission power budget used by cell c in a given time interval. Then, $\mathbf{P} = [P_1, \ldots, P_C]^\mathsf{T}$ is the transmission power budget vector for all cells. Given $\mathbf{P}$, the average SINR at user n can be modeled as follows [44]:

$$\gamma_n(\mathbf{P}) = \frac{G_{n,c(n)} P_{c(n)}}{\sigma^2 + \sum_{c \neq c(n)} G_{n,c} P_c}, \quad \forall n \in \{1, \ldots, N\}, \tag{5.18}$$

where $G_{n,c}$ denotes the channel gain between cell c and user n, $c(n)$ represents the serving cell of user n. Denote $u_n(r_n)$ as a utility function of user n when communicating at rate r_n, the power and rate control optimization problem to maximize the system's utility can be formulated as follows [44]:

$$\text{maximize} \sum_{n \in \{1, \ldots, N\}} u_n(r_n), \tag{15.19a}$$

$$\text{subject to } r_n \leq W_n \log_2(1 + \gamma_n(\mathbf{P})), \tag{15.19b}$$

$$\mathbf{P} \preccurlyeq \mathbf{P}^{\max}, \tag{15.19c}$$

where W_n is the amount of bandwidth assigned for user n and $\mathbf{P}^{\max}$ is a vector of the maximum transmission power budgets of all cells. Clearly, the above problem is nonconvex due to the inter-cell interference in the rate expression of variables $\mathbf{P}$ [44]. As such, obtaining the optimal policy by using conventional solutions could be very challenging. Moreover, the dynamics and uncertainty of the wireless system limit the applications of conventional approaches as they usually require system information in advance to formulate optimization problems. To overcome all these issues, DRL is a promising solution and has been widely adopted in the literature to effectively obtain the optimal power and rate control policies [44–58], as will be discussed in Section 5.2.2.

5.2.2 DRL-Based Power and Rate Control

In [44], the authors solve the power and rate control problem by first modeling the problem as an MDP with the state consisting of the cell power, average signal received power, average interference, and cell reward, the action is the amount of cell power, and the reward is calculated based on a utility function. Then, Q-learning algorithm is developed to obtain the optimal power and rate control policy. Simulation results demonstrated the effectiveness of the proposed solution. However, as mentioned, the Q-learning algorithm may not work well with large state and action spaces. In addition, it cannot handle continuous state and action spaces. As such, various works adopted DRL to solve the power and rate control problem. For example, in [46], the authors proposed a DRL-based approach for power control in full-duplex cognitive radio networks. In such networks, secondary users (SUs) can sense the presence of the primary users and transmit data simultaneously. Nevertheless, the sensing performance is greatly affected by the self-interference at the full-duplex transceivers. To address this issue, the authors proposed to use DRL to learn and obtain the optimal joint power control and spectrum sensing policy. In practice, the cognitive BS can only observe noisy sensing information. As such, the authors used an infinite-horizon discrete-time partially observable Markov decision process (POMDP) with a continuous state space to formulate the power control problem. The action is defined as the transmit power level for a selected SU while the reward function is the data rate per sample. The POMDP is solved by a deep Q-learning algorithm. The simulation results demonstrated that the proposed solution can outperform conventional energy detection-based sensing methods and achieve a close performance as a genie-aided method with the optimal spectrum utilization. Similarly, the joint power control and channel access problem with DRL was also investigated in [47–49], and [50].

DRL can also be applied to downlink power control. Specifically, in [51], the authors considered the power control problem in ultradense unmanned aerial vehicles (UAVs) networks. With the help of UAVs, the system capacity and network coverage can be significantly improved. However, the UAV-to-user links have additional interference and unique channel properties that are different from traditional communications. To overcome this challenge, the authors proposed to use DRL to obtain the optimal downlink power control to mitigate the interference of UAV-to-user links. To do that, the power control problem is first formulated as an MDP with the receiver interference of users and UAVs' remaining energy as the state and the UAVs' transmit power levels as the actions. Then, the deep Q-learning algorithm is developed to obtain the optimal downlink policy. Similarly, the authors in [52] and [53] proposed data-driven approaches based on DRL for the downlink power control of dense 5G networks and cell-free massive MIMO systems, respectively. Besides downlink power control, DRL has also been widely used for the uplink power control problem in [54]. In this work, DRL was adopted to obtain the optimal transmit power for the UE to maximize the data rate on the uplink as well as minimize interference caused to the neighboring cells. The simulation results indicated that by using DRL, a near-optimum performance can be achieved in terms of total transmit power, total uplink data rate, and network energy efficiency.

The authors in [55] proposed to use DRL for rate control to effectively utilize the radio links in mobile IoT networks. The authors argued that to ensure the QoS and fully utilize the uplink resources, an autonomous rate control needs to be designed for these networks. The authors first formulated the system as an MDP in which the state is a set of SNR values in previous time slots, the action is the data rate that the transceiver can choose to transmit at, and the reward function is based on the transmitted packets and the channel utilization. A deep Q-learning algorithm was then developed to solve the MDP problem to obtain the optimal rate control policy given the channel variations and uncertainties. Nevertheless, the authors in [56] pointed out that optimizing only the rate may lead to poor performance. For example, if the transmission power is decreased too much, some devices may not be able to receive the transmitted message. As such, the authors proposed a joint power and rate control optimization problem for vehicle-to-vehicle networks. The DRL agent takes the current transmit power, data rate, and estimated vehicle density as its input (i.e. system state) to learn the optimal policy, i.e. transmit power and data rate, in an online manner. The simulation results demonstrated that the proposed solution can not only alleviate congestion but also can provide enough transmission power to fulfill the coverage requirement in a vehicle-to-vehicle network.

The conventional DRL-based solutions may require a long training time to obtain the optimal policy. To speed up the training process, distributed DRL-based solutions were proposed in various studies. For example, in [57], a multi-agent DRL approach was proposed for dynamic power control in wireless networks. In the proposed approach, each transmitter in the network acts as a DRL agent. At each time slot, each agent collects CSI and QoS information from its neighbors, and then adapts its transmit power accordingly. The reward function is defined based on the weighted sum-rate of the agent and its future interfered neighbors. A deep Q-learning algorithm was then developed to obtain the optimal power control policy. However, the proposed solution in [57] only considered homogeneous networks with the assumption that neighboring transceivers (i.e. AP and users) can exchange their information. This limits its applications in practice. As such, the authors in [58] proposed a novel multi-agent DRL-based approach that only requires local information for heterogeneous networks (HetNets). In particular, the authors considered a typical HetNet in which multiple APs share the same spectrum band to serve several users that may cause interference to each other (as shown in Figure 5.4). To avoid exchanging information between agents (i.e. APs), the authors proposed a multiagent DDPG algorithm with a centralized-training-distributed-execution framework. Specifically, each AP has its own local deep neural network (DNN), and the local DNNs are trained

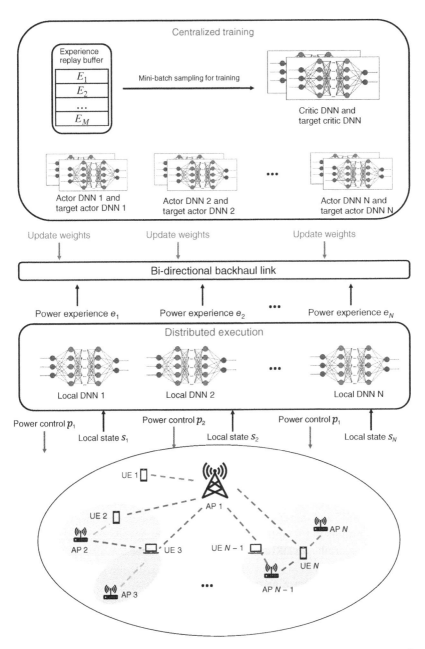

Figure 5.4 A general HetNet, in which multiple APs share the same spectrum band to serve the users within their coverage and may cause interference to each other. The figure is adapted from [58].

separately with multiple actors and one critic network as shown in Figure 5.4. In this way, the actor network can be used to determine whether the policy (i.e. transmit power) of each actor is good or not from a global view. The simulation results then demonstrated that the proposed solution outperforms the conventional power control algorithms in terms of average sum-rate and complexity.

5.3 Physical-Layer Security

Wireless communications are extremely vulnerable to security threats due to their open medium nature. One of the most dangerous types of security threats is jamming attacks. To perform jamming attacks, a jammer can intentionally generate and send strong interference signals to the target system to disrupt communications of the legitimate devices. Dealing with jamming attacks, however, can be very challenging depending on the dynamics of the wireless environment and intelligence of the jammer. As such, conventional approaches may not be able to mitigate the jamming attacks efficiently. On the other hand, DRL has been attracting more attention from researchers as a promising solution to deal with jamming attacks [59–66]. For example, the authors in [59] proposed a two-dimensional anti-jamming communication scheme for cognitive communication networks based on DRL. In particular, the proposed scheme can obtain the optimal defense policy to guide the SU to leave an area of heavy jamming and select a frequency hopping (FH) pattern to communicate. To do that, the anti-jamming problem was formulated as a Markov game in which the SU tries to decide whether it moves to another area and selects a different FH pattern, and the jammer tries to find a channel to attack. This game was then solved by a deep Q-learning algorithm to obtain the optimal defense policy for the SU. The simulation results demonstrated that by using DRL, the authors can achieve better performance than other schemes in the literature. However, in the proposed solution, the jammer is assumed to have the same channel-slot transmission structure as the SU, and therefore limiting its applications in practice. To solve this problem, the authors in [61] introduced an anti-jamming approach that leverages the spectrum waterfall and DRL. In this way, the estimated jamming patterns and parameters are not required in the learning process while the proposed solution still achieves good anti-jamming performance.

Although the above approaches achieve good performance in dealing with jamming attacks, since they are mostly based on FH and rate adaptation (RA) techniques, they may not work well if the jammer can attack multiple channels simultaneously with very high-power levels. To deal with this issue, the authors in [64] proposed a novel anti-jamming approach based on ambient backscatter communications and DRL. With ambient backscatter communication technology, the transmitter can send its information to the receiver by backscattering the jamming signals, allowing it to operate under jamming attacks. The authors demonstrated that with higher jamming power, the transmitter can backscatter more data to the receiver. DRL is then used to learn the jammer's attack strategy and obtain the optimal anti-jamming policy. The authors then extended this solution to deal with a more intelligent jammer in [65]. Specifically, the authors aimed at dealing with reactive jamming attacks in which the jammer can sense the channel and attack the channel if it detects communications from the legitimate transmitter. For this reason, the solution proposed in [64] is no longer effective. The authors then proposed a deception mechanism to trap the reactive jammer as illustrated in Figure 5.5. In particular, the transmitter can transmit "fake" signals to attract the jammer and then leverage the jamming signals when the jammer attacks the channel. The deep dueling algorithm, an advanced DRL approach, is then developed to quickly obtain the optimal defense policy, i.e. when to perform the deception, when to backscatter, or when to stay idle, for the transmitter. The simulation results confirmed the effectiveness of the proposed deception mechanism in dealing with reactive jamming attacks.

Recently, RISs have been widely adopted to secure wireless networks. In [67], the authors proposed to use an RIS to enhance anti-jamming communication performance and mitigate jamming interference by adjusting the surface reflecting elements of the RIS. To do that, the authors first formulated the problem as an MDP in which the state includes the previous jamming power and the estimated channel coefficients, the action consists of the transmit power of the BS and the reflecting beamforming coefficient. The optimal policy was then obtained by using a fuzzy win learning algorithm, a fast reinforcement-learning algorithm. Through the simulation results, the authors demonstrated that the proposed solution can not only improve the transmission protection level but also increase the system rate compared to the existing solutions.

Wireless networks are vulnerable to passive attacks, e.g. those launched by the eavesdroppers. The authors in [11] focused on the anti-eavesdropper problem by leveraging RIS and DRL. Specifically, an RIS can be used to degrade the reception quality at the eavesdropper by adjusting its elements. A DRL method was proposed to obtain the joint optimal beamforming policy for both the BS and the RIS. To further enhance the transmission protection level and quickly obtain the optimal beamforming policy, the authors in [68] developed a multi-agent DRL approach for intelligent reflecting surface (IRS)-assisted secure cooperative networks.

5.4 Chapter Summary

In this chapter, we have discussed many applications of DRL to the physical layer of wireless communications systems. Specifically, we have first discussed the applications of DRL for beamforming, channel estimation, and decoding. While

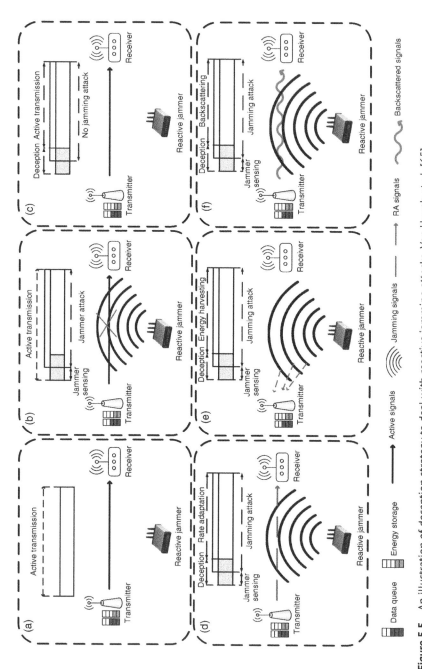

Figure 5.5 An illustration of deception strategy to deal with reactive jamming attacks. Van Huynh et al. [65].

there are many papers working on beamforming problems, we observe that the applications of DRL for channel estimation and decoding are rather limited. Then, we have reviewed various DRL-based approaches for the power and rate control problem. Finally, solutions for securing wireless networks at the physical layer (i.e. physical layer security methods) based on DRL have been highlighted.

References

1 F. B. Mismar, B. L. Evans, and A. Alkhateeb, "Deep reinforcement learning for 5G networks: Joint beamforming, power control, and interference coordination," *IEEE Transactions on Communications*, vol. 68, no. 3, pp. 1581–1592, 2019.

2 J. Ge, Y.-C. Liang, J. Joung, and S. Sun, "Deep reinforcement learning for distributed dynamic MISO downlink-beamforming coordination," *IEEE Transactions on Communications*, vol. 68, no. 10, pp. 6070–6085, 2020.

3 J. Zhang, H. Zhang, Z. Zhang, H. Dai, W. Wu, and B. Wang, "Deep reinforcement learning-empowered beamforming design for IRS-assisted MISO interference channels," in *2021 13th International Conference on Wireless Communications and Signal Processing (WCSP)*, pp. 1–5, IEEE, 2021.

4 J. Kim, S. Hosseinalipour, T. Kim, D. J. Love, and C. G. Brinton, "Multi-IRS-assisted multi-cell uplink MIMO communications under imperfect CSI: A deep reinforcement learning approach," in *2021 IEEE International Conference on Communications Workshops (ICC Workshops)*, pp. 1–7, IEEE, 2021.

5 H. Chen, Z. Zheng, X. Liang, Y. Liu, and Y. Zhao, "Beamforming in multi-user MISO cellular networks with deep reinforcement learning," in *2021 IEEE 93rd Vehicular Technology Conference (VTC2021-Spring)*, pp. 1–5, IEEE, 2021.

6 Y. Lee, J.-H. Lee, and Y.-C. Ko, "Beamforming optimization for IRS-assisted mmWave V2I communication systems via reinforcement learning," *IEEE Access*, vol. 10, pp. 60521–60533, 2022.

7 H. Jia, Z.-Q. He, H. Rui, and W. Lin, "Robust distributed MISO beamforming using multi-agent deep reinforcement learning," in *2022 14th International Conference on Communication Software and Networks (ICCSN)*, pp. 197–201, IEEE, 2022.

8 F. Fredj, Y. Al-Eryani, S. Maghsudi, M. Akrout, and E. Hossain, "Distributed beamforming techniques for cell-free wireless networks using deep reinforcement learning," *IEEE Transactions on Cognitive Communications and Networking*, vol. 8, no. 2, pp. 1186–1201, 2022.

9 C. Huang, Z. Yang, G. C. Alexandropoulos, K. Xiong, L. Wei, C. Yuen, and Z. Zhang, "Hybrid beamforming for RIS-empowered multi-hop terahertz communications: A DRL-based method," in *2020 IEEE Globecom Workshops (GC Wkshps)*, pp. 1–6, IEEE, 2020.

10 J. Lin, Y. Zout, X. Dong, S. Gong, D. T. Hoang, and D. Niyato, "Deep reinforcement learning for robust beamforming in IRS-assisted wireless communications," in *GLOBECOM 2020-2020 IEEE Global Communications Conference*, pp. 1–6, IEEE, 2020.

11 H. Yang, Z. Xiong, J. Zhao, D. Niyato, L. Xiao, and Q. Wu, "Deep reinforcement learning-based intelligent reflecting surface for secure wireless communications," *IEEE Transactions on Wireless Communications*, vol. 20, no. 1, pp. 375–388, 2020.

12 X. Pang, N. Zhao, J. Tang, C. Wu, D. Niyato, and K.-K. Wong, "IRS-assisted secure UAV transmission via joint trajectory and beamforming design," *IEEE Transactions on Communications*, vol. 70, no. 2, pp. 1140–1152, 2021.

13 E. M. Lizarraga, G. N. Maggio, and A. A. Dowhuszko, "Hybrid beamforming algorithm using reinforcement learning for millimeter wave wireless systems," in *2019 XVIII Workshop on Information Processing and Control (RPIC)*, pp. 253–258, IEEE, 2019.

14 Q. Wang, K. Feng, X. Li, and S. Jin, "PrecoderNet: Hybrid beamforming for millimeter wave systems with deep reinforcement learning," *IEEE Wireless Communications Letters*, vol. 9, no. 10, pp. 1677–1681, 2020.

15 E. M. Lizarraga, G. N. Maggio, and A. A. Dowhuszko, "Deep reinforcement learning for hybrid beamforming in multi-user millimeter wave wireless systems," in *2021 IEEE 93rd Vehicular Technology Conference (VTC2021-Spring)*, pp. 1–5, IEEE, 2021.

16 Q. Wang, X. Li, S. Jin, and Y. Chen, "Hybrid beamforming for mmWave MU-MISO systems exploiting multi-agent deep reinforcement learning," *IEEE Wireless Communications Letters*, vol. 10, no. 5, pp. 1046–1050, 2021.

17 M. Fozi, A. R. Sharafat, and M. Bennis, "Fast MIMO beamforming via deep reinforcement learning for high mobility mmWave connectivity," *IEEE Journal on Selected Areas in Communications*, vol. 40, no. 1, pp. 127–142, 2021.

18 X. Zhou, X. Zhang, C. Chen, Y. Niu, Z. Han, H. Wang, C. Sun, B. Ai, and N. Wang, "Deep reinforcement learning coordinated receiver beamforming for millimeter-wave train-ground communications," *IEEE Transactions on Vehicular Technology*, vol. 71, no. 5, pp. 5156–5171, 2022.

19 H. Vaezy, M. S. H. Abad, O. Ercetin, H. Yanikomeroglu, M. J. Omidi, and M. M. Naghsh, "Beamforming for maximal coverage in mmWave drones: A reinforcement learning approach," *IEEE Communications Letters*, vol. 24, no. 5, pp. 1033–1037, 2020.

20 M. Liu, R. Wang, Z. Xing, and I. Soto, "Deep reinforcement learning based dynamic power and beamforming design for time-varying wireless downlink interference channel," in *2022 IEEE Wireless Communications and Networking Conference (WCNC)*, pp. 471–476, IEEE, 2022.

21 J. Liu, C.-H. R. Lin, Y.-C. Hu, and P. K. Donta, "Joint beamforming, power allocation, and splitting control for SWIPT-enabled IoT networks with deep reinforcement learning and game theory," *Sensors*, vol. 22, no. 6, p. 2328, 2022.

22 C. Wang, C.-L. Hsieh, and H.-C. Gao, "Deep reinforcement-learning based power and beamforming coordination for IoT underlying 5G networks," in *2021 International Conference on Electronic Communications, Internet of Things and Big Data (ICEIB)*, pp. 20–22, IEEE, 2021.

23 A. Akbarpour-Kasgari and M. Ardebilipour, "Joint power allocation and beamformer for mmW-NOMA downlink systems by deep reinforcement learning," *arXiv preprint arXiv:2205.06489*, 2022.

24 A. Özçelikkale, M. Koseoglu, M. Srivastava, and A. Ahlén, "Deep reinforcement learning based energy beamforming for powering sensor networks," in *2019 IEEE 29th International Workshop on Machine Learning for Signal Processing (MLSP)*, pp. 1–6, IEEE, 2019.

25 T. Lu, H. Zhang, and K. Long, "Joint beamforming and power control for MIMO-NOMA with deep reinforcement learning," in *ICC 2021-IEEE International Conference on Communications*, pp. 1–5, IEEE, 2021.

26 K. Ma, Z. Wang, W. Tian, S. Chen, and L. Hanzo, "Deep learning for mmWave beam-management: State-of-the-art, opportunities and challenges," *IEEE Wireless Communications*, pp. 1–8, 2022.

27 T.-K. Kim, Y.-S. Jeon, J. Li, N. Tavangaran, and H. V. Poor, "Semi-data-aided channel estimation for MIMO systems via reinforcement learning," *arXiv preprint arXiv:2204.01052*, 2022.

28 Y.-S. Jeon, J. Li, N. Tavangaran, and H. V. Poor, "Data-aided channel estimator for MIMO systems via reinforcement learning," in *ICC 2020-2020 IEEE International Conference on Communications (ICC)*, pp. 1–6, IEEE, 2020.

29 M. S. Oh, S. Hosseinalipour, T. Kim, C. G. Brinton, and D. J. Love, "Channel estimation via successive denoising in MIMO OFDM systems: A reinforcement learning approach," in *ICC 2021-IEEE International Conference on Communications*, pp. 1–6, IEEE, 2021.

30 X. Li, Q. Wang, H. Yang, and X. Ma, "Data-aided MIMO channel estimation by clustering and reinforcement-learning," in *2022 IEEE Wireless Communications and Networking Conference (WCNC)*, pp. 584–589, IEEE, 2022.

31 Y.-S. Jeon, N. Lee, and H. V. Poor, "Robust data detection for MIMO systems with one-bit ADCs: A reinforcement learning approach," *IEEE Transactions on Wireless Communications*, vol. 19, no. 3, pp. 1663–1676, 2019.

32 Y.-S. Jeon, M. So, and N. Lee, "Reinforcement-learning-aided ML detector for uplink massive MIMO systems with low-precision ADCs," in *2018 IEEE Wireless Communications and Networking Conference (WCNC)*, pp. 1–6, IEEE, 2018.

33 T.-K. Kim, Y.-S. Jeon, and M. Min, "Training length adaptation for reinforcement learning-based detection in time-varying massive MIMO systems with one-bit ADCs," *IEEE Transactions on Vehicular Technology*, vol. 70, no. 7, pp. 6999–7011, 2021.

34 M. Chu, A. Liu, V. K. Lau, C. Jiang, and T. Yang, "Deep reinforcement learning based end-to-end multi-user channel prediction and beamforming," *IEEE Transactions on Wireless Communications*, vol. 21, no. 12, pp. 10271–10285, 2022.

35 T.-K. Kim and M. Min, "A low-complexity algorithm for a reinforcement learning-based channel estimator for MIMO systems," *Sensors*, vol. 22, no. 12, p. 4379, 2022.

36 F. Carpi, C. Häger, M. Martalò, R. Raheli, and H. D. Pfister, "Reinforcement learning for channel coding: Learned bit-flipping decoding," in *2019 57th Annual Allerton Conference on Communication, Control, and Computing (Allerton)*, pp. 922–929, IEEE, 2019.

37 N. Doan, S. A. Hashemi, F. Ercan, and W. J. Gross, "Fast SC-flip decoding of polar codes with reinforcement learning," in *ICC 2021-IEEE International Conference on Communications*, pp. 1–6, IEEE, 2021.

38 Y. Liao, S. A. Hashemi, J. M. Cioffi, and A. Goldsmith, "Construction of polar codes with reinforcement learning," *IEEE Transactions on Communications*, vol. 70, no. 1, pp. 185–198, 2021.

39 N. Doan, S. A. Hashemi, and W. J. Gross, "Decoding polar codes with reinforcement learning," in *GLOBECOM 2020-2020 IEEE Global Communications Conference*, pp. 1–6, IEEE, 2020.

40 J. Gao and K. Niu, "A reinforcement learning based decoding method of short polar codes," in *2021 IEEE Wireless Communications and Networking Conference Workshops (WCNCW)*, pp. 1–6, IEEE, 2021.

41 S. Habib, A. Beemer, and J. Kliewer, "Learning to decode: Reinforcement learning for decoding of sparse graph-based channel codes," *Advances in Neural Information Processing Systems 33*, pp. 22396–22406, 2020.

42 S. Habib, A. Beemer, and J. Kliewer, "RELDEC: Reinforcement learning-based decoding of moderate length LDPC codes," *arXiv preprint arXiv:2112.13934*, 2021.

43 S. Habib, A. Beemer, and J. Kliewer, "Belief propagation decoding of short graph-based channel codes via reinforcement learning," *IEEE Journal on Selected Areas in Information Theory*, vol. 2, no. 2, pp. 627–640, 2021.

44 E. Ghadimi, F. D. Calabrese, G. Peters, and P. Soldati, "A reinforcement learning approach to power control and rate adaptation in cellular networks," in *2017 IEEE International Conference on Communications (ICC)*, pp. 1–7, IEEE, 2017.

45 P. Zhou, W. Yuan, W. Liu, and W. Cheng, "Joint power and rate control in cognitive radio networks: A game-theoretical approach," in *2008 IEEE International Conference on Communications*, pp. 3296–3301, IEEE, 2008.

46 X. Meng, H. Inaltekin, and B. Krongold, "Deep reinforcement learning-based power control in full-duplex cognitive radio networks," in *2018 IEEE Global Communications Conference (GLOBECOM)*, pp. 1–7, IEEE, 2018.

47 Z. Lu and M. C. Gursoy, "Dynamic channel access and power control via deep reinforcement learning," in *2019 IEEE 90th Vehicular Technology Conference (VTC2019-Fall)*, pp. 1–5, IEEE, 2019.

48 H. Zhang, N. Yang, W. Huangfu, K. Long, and V. C. Leung, "Power control based on deep reinforcement learning for spectrum sharing," *IEEE Transactions on Wireless Communications*, vol. 19, no. 6, pp. 4209–4219, 2020.

49 J. Tan, Y.-C. Liang, L. Zhang, and G. Feng, "Deep reinforcement learning for joint channel selection and power control in D2D networks," *IEEE Transactions on Wireless Communications*, vol. 20, no. 2, pp. 1363–1378, 2020.

50 G. Zhao, Y. Li, C. Xu, Z. Han, Y. Xing, and S. Yu, "Joint power control and channel allocation for interference mitigation based on reinforcement learning," *IEEE Access*, vol. 7, pp. 177254–177265, 2019.

51 L. Li, Q. Cheng, K. Xue, C. Yang, and Z. Han, "Downlink transmit power control in ultra-dense UAV network based on mean field game and deep reinforcement learning," *IEEE Transactions on Vehicular Technology*, vol. 69, no. 12, pp. 15594–15605, 2020.

52 S. Saeidian, S. Tayamon, and E. Ghadimi, "Downlink power control in dense 5G radio access networks through deep reinforcement learning," in *ICC 2020-2020 IEEE International Conference on Communications (ICC)*, pp. 1–6, IEEE, 2020.

53 L. Luo, J. Zhang, S. Chen, X. Zhang, B. Ai, and D. W. K. Ng, "Downlink power control for cell-free massive MIMO with deep reinforcement learning," *IEEE Transactions on Vehicular Technology*, vol. 71, no. 6, pp. 6772–6777, 2022.

54 F. H. C. Neto, D. C. Araújo, M. P. Mota, T. F. Maciel, and A. L. de Almeida, "Uplink power control framework based on reinforcement learning for 5G networks," *IEEE Transactions on Vehicular Technology*, vol. 70, no. 6, pp. 5734–5748, 2021.

55 W. Xu, H. Zhou, N. Cheng, N. Lu, L. Xu, M. Qin, and S. Guo, "Autonomous rate control for mobile Internet of Things: A deep reinforcement learning approach," in *2020 IEEE 92nd Vehicular Technology Conference (VTC2020-Fall)*, pp. 1–6, IEEE, 2020.

56 J. Aznar-Poveda, A.-J. Garcia-Sanchez, E. Egea-Lopez, and J. García-Haro, "Simultaneous data rate and transmission power adaptation in V2V communications: A deep reinforcement learning approach," *IEEE Access*, vol. 9, pp. 122067–122081, 2021.

57 Y. S. Nasir and D. Guo, "Multi-agent deep reinforcement learning for dynamic power allocation in wireless networks," *IEEE Journal on Selected Areas in Communications*, vol. 37, no. 10, pp. 2239–2250, 2019.

58 L. Zhang and Y.-C. Liang, "Deep reinforcement learning for multi-agent power control in heterogeneous networks," *IEEE Transactions on Wireless Communications*, vol. 20, no. 4, pp. 2551–2564, 2020.

59 G. Han, L. Xiao, and H. V. Poor, "Two-dimensional anti-jamming communication based on deep reinforcement learning," in *2017 IEEE International Conference on Acoustics, Speech and Signal Processing (ICASSP)*, pp. 2087–2091, IEEE, 2017.

60 S. Liu, Y. Xu, X. Chen, X. Wang, M. Wang, W. Li, Y. Li, and Y. Xu, "Pattern-aware intelligent anti-jamming communication: A sequential deep reinforcement learning approach," *IEEE Access*, vol. 7, pp. 169204–169216, 2019.

61 X. Liu, Y. Xu, L. Jia, Q. Wu, and A. Anpalagan, "Anti-jamming communications using spectrum waterfall: A deep reinforcement learning approach," *IEEE Communications Letters*, vol. 22, no. 5, pp. 998–1001, 2018.

62 X. Lu, L. Xiao, C. Dai, and H. Dai, "UAV-aided cellular communications with deep reinforcement learning against jamming," *IEEE Wireless Communications*, vol. 27, no. 4, pp. 48–53, 2020.

63 C. Han, L. Huo, X. Tong, H. Wang, and X. Liu, "Spatial anti-jamming scheme for internet of satellites based on the deep reinforcement learning and stackelberg game," *IEEE Transactions on Vehicular Technology*, vol. 69, no. 5, pp. 5331–5342, 2020.

64 N. Van Huynh, D. N. Nguyen, D. T. Hoang, and E. Dutkiewicz, ""Jam me if you can:" Defeating jammer with deep dueling neural network architecture and ambient backscattering augmented communications," *IEEE Journal on Selected Areas in Communications*, vol. 37, no. 11, pp. 2603–2620, 2019.

65 N. Van Huynh, D. T. Hoang, D. N. Nguyen, and E. Dutkiewicz, "DeepFake: Deep dueling-based deception strategy to defeat reactive jammers," *IEEE Transactions on Wireless Communications*, vol. 20, no. 10, pp. 6898–6914, 2021.

66 N. Van Huynh, D. N. Nguyen, D. T. Hoang, T. X. Vu, E. Dutkiewicz, and S. Chatzinotas, "Defeating super-reactive jammers with deception strategy: Modeling, signal detection, and performance analysis," *IEEE Transactions on Wireless Communications*, vol. 21, no. 9, pp. 7374–7390, 2022.

67 H. Yang, Z. Xiong, J. Zhao, D. Niyato, Q. Wu, H. V. Poor, and M. Tornatore, "Intelligent reflecting surface assisted anti-jamming communications: A fast reinforcement learning approach," *IEEE transactions on wireless communications*, vol. 20, no. 3, pp. 1963–1974, 2020.

68 C. Huang, G. Chen, and K.-K. Wong, "Multi-agent reinforcement learning-based buffer-aided relay selection in IRS-assisted secure cooperative networks," *IEEE Transactions on Information Forensics and Security*, vol. 16, pp. 4101–4112, 2021.

6

DRL at the MAC Layer

6.1 Resource Management and Optimization

Due to the ever-increasing demands of wireless and mobile communications services, wireless systems have continuously evolved to support new applications and a diversity of traffic types. For example, while the 4G systems mainly focus on improving throughput, the current 5G systems (and beyond) are considered use-case-driven technologies [1] which will also focus on reliability and latency metrics. Currently, the primary use-cases supported by 5G are ultra-reliable and low latency communications (URLLC), enhanced mobile broadband (eMBB), and massive machine type communications (mMTC). Each of these use cases comes with different service requirements on delay, latency, capacity, and mobility. In the future, more use cases are expected to emerge to support a massive number of ongoing wireless devices with heterogeneous quality of service (QoS) requirements [2]. Even with advanced network resource slicing, the resource optimization problem (e.g. resource blocks [RBs]) at the medium access control (MAC) layer can become challenging, making conventional approaches inherited from previous network generations unsuitable [3].

In this context, deep reinforcement learning (DRL) is a promising approach to address the dynamics and scalability challenges at the MAC layer. First, using DRL at the MAC layer allows an agent (e.g. a mobile device or a base station [BS]) to learn an optimal policy via interactions with the environment instead of requiring a large data set in advance for the learning process (as in other types of machine learning [ML] approaches such as the supervised learning approaches). Second, by using a deep neural network (DNN), DRL effectively addresses the curse-of-dimensionality problem of conventional approaches using the optimization theory. Third, the underlying Markov decision process (MDP) of DRL allows an agent (e.g. a base station and user equipment) adaptively make the best decision to maximize the long-term system performance without requiring complete information of the surrounding environment.

Deep Reinforcement Learning for Wireless Communications and Networking:
Theory, Applications, and Implementation, First Edition.
Dinh Thai Hoang, Nguyen Van Huynh, Diep N. Nguyen, Ekram Hossain, and Dusit Niyato.
© 2023 The Institute of Electrical and Electronics Engineers, Inc. Published 2023 by John Wiley & Sons, Inc.

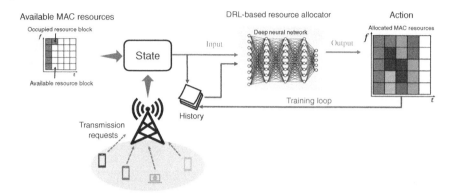

Figure 6.1 DRL-based model for MAC resource optimization.

DRL-based resource management at the MAC layer can be illustrated in Figure 6.1. A DRL-based resource allocator (i.e. an agent) is typically implemented at a BS/access point. The agent observes available resources (e.g. frequency and time) together with transmission demands from end-users to determine the resources allocated to each user. After performing an action, by observing performance metrics such as total system throughput and fairness, the agent can adjust its current policy to gradually learn/converge to an optimal policy π to maximize the system performance. At time t, we can define a state s_t as an observation of current system resources, and action a_t as a resource allocation decision. The immediate reward r_t for taking action a_t should reflect the system performance, e.g. throughput, fairness, energy efficiency, or reliability. Thus, the resource optimization problem can be formulated as follows:

$$\max_{\pi} R(\pi) = \lim_{T \to \infty} \frac{1}{T} \sum_{t=1}^{T} \mathbb{E}\left(r_t(s_t, \pi(s_t))\right), \tag{6.1}$$

where $R(\pi)$ is the long-term average reward functions and $\pi(s_t) = a_t$.

In the literature, several works investigated different aspects of resource optimization at the MAC layer, e.g. [1, 4]. In [1], the authors studied a radio resource scheduling problem in the MAC layer of the 5G network. They consider a scenario in which U user equipments (UEs) connect to a gNodeB, i.e. a 3GPP-defined 5G BS. A radio resource scheduler (RRS) implemented at the gNodeB is responsible for allocating available RBs (in the frequency-time grid) to active UEs. In this work, the system state is defined by combining three components. The first one is a vector describing whether a UE is eligible to be allocated resources. The second component is a data rate vector whose each element captures the data rate required by the corresponding UE. The third component is a fairness vector reflecting the allocation-log of the resources. The action to select which UE would be allocated RBs is represented by a vector whose all elements are zero except the one corresponding to the selected UE. The immediate reward is defined as a function of data rate and fairness. To overcome the intractability of conventional optimization approaches in the 5G MAC layer, the authors proposed a scheduler based on the Deep Double Q-learning algorithm [5], namely, LEASCH. The simulation results show that the proposed scheduler outperforms conventional solutions in terms of throughput, goodput, and fairness.

Similarly, the study in [4] also considered the problem of radio resource allocation in the 5G MAC layer but for two use-cases, namely, eMBB and URLLC. In particular, the authors studied the multiplexing of URLLC and eMBB on the same channel at a gNodeB. After a URLLC packet arrives at the BS, it will be scheduled to be transmitted in the next mini-time slot to fulfill its delay requirement. If all the subcarriers (SCs) are occupied by ongoing eMBB traffic, the BS will puncture/preempt a certain amount of resources from eMBB transmissions to allocate for URLLC. If too much resources are taken from one eMBB user, the service quality for this user will be severely degraded. The objective of this work is to find the optimal amount of resources to be preempted from each eMBB user so that the impact on eMBB users is minimized while guaranteeing the service requirements for URLLC users. Since the arrival process of URLLC packets is uncertain and often unknown in advance, the authors proposed a solution, namely DEMUX, based on deep deterministic policy gradient (DDPG) [6] method. In this study, a state consists of two vectors: (i) the SC allocation vector whose each element captures resources that are being occupied by the corresponding eMBB transmission, and (ii) the data rate vector whose each element represents the modulation and coding scheme (MCS) selection of the corresponding eMBB transmission. An action of the gNodeB is represented by a vector, each element of which indicates the percentage of resources taken from the corresponding eMBB transmission. After taking an action, the agent receives an immediate reward based on an impact of this action on the eMBB traffic and a penalty for any infeasible action at a state. The simulation results show that DEMUX can achieve a higher throughput than the algorithm proposed in the 3GPP standard while guaranteeing the real-time requirement of 5G scheduling.

6.2 Channel Access Control

With a reservation-based MAC approach, since RBs are allocated to users exclusively, there will be no collision between users' transmission. However, in practice, due to the high dynamics and uncertainty of users' data transmission, such an approach may lead to under-utilization of network resources. This issue can be addressed by a dynamic channel access control mechanism, which will allow users to access channels without precoordination but based on their demand. In this case, collisions can happen if two or more users transmit data on the same channel at the same time. Thus, channel access control requires each user to follow a procedure to access a shared channel, e.g. the listen-before-talk (LBT) protocol in which users first listen/sense a channel and then transmit data only if the channel is idle for a given predetermined duration.

Given the above context, as illustrated in Figure 6.2, DRL-based channel access controllers deployed on user devices can act as learning agents. At time slot t, the state s_t is typically defined as a channel status history (e.g. idle, collision, and successful transmission) and/or collision probability. The action a_t is to whether transmit data or wait at time slot t after observing state s_t. The reward r_t of taking an action a_t captures the efficiency of the channel access policy that can be either the number of successful transmissions or the system's throughput. DRL-based algorithms can help channel access controllers find their optimal access policies that maximize a long-term average reward function, which can be formulated similar to Eq. (6.1), i.e.

$$\pi^* = \operatorname*{argmax}_{\pi} R(\pi) = \operatorname*{argmax}_{\pi} \left[\lim_{T \to \infty} \frac{1}{T} \sum_{t=1}^{T} \mathbb{E}\big(r_t(s_t, \pi(s_t))\big) \right]. \tag{6.2}$$

This section will discuss the state-of-the-art DRL-based approaches for channel access control in different systems, including the IEEE 802.11, IoT, and 5G systems.

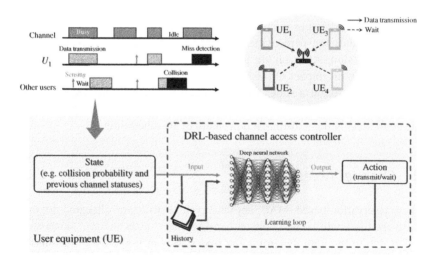

Figure 6.2 DRL-based channel access control.

6.2.1 DRL in the IEEE 802.11 MAC

In the IEEE 802.11 standard, channel access control is based on the carrier-sense multiple access with collision avoidance (CSMA/CA) method that relies on the back-off procedure [7]. Specifically, when a device wants to transmit data, it first senses the channel to determine whether it is idle or busy. Then, if the channel is idle, the device needs to "back-off", i.e. wait for a certain number of time slots before transmitting data. Typically, the number of time slots is randomly sampled from the contention window (CW) value range, which is often statically configured. Therefore, it may lead to an inefficient operation when the surrounding environment changes, e.g. the number of devices. Several works aimed to study the back-off procedure in IEEE 802.11 MAC [8, 9]. In [8], the authors aim to find the optimal CW value for the channel access procedure of 802.11ax. The authors consider a model in which multiple devices are served by an access point (AP). Since it is challenging to obtain the status of all devices connected to the AP, they formulate the problem as a partially observable Markov decision process (POMDP), where an agent (i.e. the AP) only observes partial information of the network status to determine the value of CW (i.e. an action). This paper defines an observation of the agent as a collision probability. After taking an action (i.e. the value of CW), the agent receives a reward that captures the network throughput. To find an optimal value of CW, the authors developed DRL-based algorithms using the deep Q-learning (DQN) [10] and DDPG mechanisms. The simulation results show that their proposed solutions achieve similar results but both outperform the standard CW control defined in the IEEE 802.11ax.

Similar to [8], the work in [9] also investigated the MAC mechanism for IEEE 802.11ax. In particular, the authors studied the problem of resource unit (RU) selection in the multiuser orthogonal frequency division multiple access (MU-OFDMA) transmission cycle of the IEEE 802.11ax [11]. The authors consider a network consisting of one AP (namely STAs) serving multiple wireless devices. At the beginning of an MU-OFDMA transmission cycle, the AP broadcasts a trigger frame (TR-F) that provides information about the network, such as available RUs and frame duration. Upon receiving the TR-F, STAs decide when to transmit their data based on the OFDMA back-off procedure, where each STA selects the waiting (i.e. back-off) time. After that, it randomly selects one of the available RUs to send its data. Thus, the randomness in selecting the RU may result in a high probability of collision among STAs, leading to a decrease in throughput and an increase in latency. To reduce collisions, the authors proposed a method based on the DQN algorithm [10], namely, C-DRL, that is deployed at each STA. In this work, a state is defined as the status of all RUs (i.e. idle, successful transmission, and collision) over M previously participated transmissions. For N RUs, a state can be represented as an image with $N \times M$ size. The reward is defined as a discounted sum of the number of successful transmissions in the M previous observations. To speed up the learning process, they propose a greedy experience replay mechanism to provide synthetic experiences that is created "artificially" based on the observation of RUs' status instead of real interactions with the surrounding environment. Each synthetic experience is a tuple of a current state, action, reward, and a next state. Their simulation results show that the proposed solution achieves a greater throughput and lower latency compared with those of the standard access control in IEEE 802.11ax.

Unlike the above works (i.e. [8, 9]), a channel access problem at the MAC layer for dense wireless local area networks (WLANs) was investigated in [12]. At the 802.11 MAC layer, before accessing a channel, all STAs have to follow the back-off mechanism to avoid collision with others [7]. Specifically, these devices need to wait for a period of time randomly and independently generated from a range of CW values, that double at every collision or reset to a default value (i.e. minimum CW value) after a successful transmission. However, the reset of CW may lead to a higher collision probability in dense networks, thereby degrading the network performance. To handle this problem, the authors adopt a channel observation-based scaled back-off (COSB) mechanism. The performance of COSB is then optimized by leveraging the Q-learning algorithm. In the MDP formulation, the state is defined as the stage of back-off, i.e. CW size. An agent, i.e. an STA, observes a state and decides whether to increase or decrease the CW value based on COSB. With an objective of minimizing the collision probability p_{obs}, a reward for performing an action is defined as $1 - p_{obs}$. The simulation results show that the proposed approach can outperform the standard mechanism in terms of throughput while maintaining a comparable fairness level.

While the aforementioned studies (i.e. [8, 9, 12]) attempted to replace traditional MAC protocol with DRL-based protocols that can automatically and smartly tune or control the protocol parameters (e.g. when to access the channel and), the work in [13] focused on optimizing functionalities in designing an MAC protocol for WLANs. In particular, the authors propose a DRL-based framework for designing IEEE 802.11 MAC protocol, namely, DeepMAC, to decompose a protocol into multiple parametric modules, namely building blocks (BBs). Each BB represents one or more protocol's functionality. For example, the distributed coordination function (DCF) (i.e. the fundamental MAC technique of the IEEE 802.11-based standard) can be decomposed into multiple BBs, such as ACK, contention window control, and carrier sensing. The combination of BBs and their interactions determines the protocol's behavior. The main idea of DeepMAC is that given the high-level specification of a scenario (e.g. BBs of the protocol, network configuration, and communication objective), the DeepMAC can determine which BBs are necessary or can be neglected in designing a protocol for this particular scheme. The authors adopt DRL to find an optimal policy (i.e. the best set of BBs) for a specific scenario. At a time slot, the DeepMAC agent deployed on an AP observes the state (i.e. a numerical vector consisting of BBs and history with a fixed length of the average throughput) to determine which blocks will be used for the MAC protocol (i.e. an action). After taking an action, the agent receives an immediate reward, which is the average throughput during this time slot. The simulation results show that the proposed solution can adjust the MAC protocol according to the changes in simulation scenarios.

6.2.2 MAC for Massive Access in IoT

Unlike the IEEE 802.11-based WLAN, where the number of concurrently connected devices is not very large, an Internet of Things (IoT) network may have thousands of devices, each with different data transmission intervals and limited resources (e.g. computing and energy). Thus, channel access control approaches for IoT need to address the massive access problem while keeping the complexity at a reasonable level. In addition, the access demand in IoT networks is not only much greater than that of conventional WLANs but also highly dynamic and uncertain, which makes designing the MAC protocols more challenging. In this context, DRL can be used to tackle emerging challenges in designing IoT MAC protocols [14–17].

In IoT networks, due to the nature of irregular communication and the requirement of channel utilization efficiency, it is preferable to use probabilistic MAC protocols (e.g. *p*-persistent slotted ALOHA [7]) over deterministic MAC protocols, e.g. time-division multiple access (TDMA). However, probabilistic MAC protocols are prone to collisions when two or more devices access a shared channel simultaneously. This problem becomes more severe in an IoT network consisting of a very large number of devices. To this end, as illustrated in Figure 6.3, several DRL frameworks were proposed to address this problem of probabilistic MAC protocols [14–16]. In [14], the authors studied an MAC problem for wireless sensor networks (WSNs) that consists of multiple devices sensing the surrounding environment and sending collected data to a core network via a shared channel. In particular, the considered WSN adopts an irregular repetition slotted ALOHA (IRSA) method that can address the collision problem by leveraging successive interference cancelation (SIC) [18]. The main idea of SIC is that when two or more packets arrive at a receiver simultaneously, the device first decodes the stronger signal and then subtracts it from the combined signal before decoding weaker signals. To achieve the time-domain diversity in IRSA, devices send multiple replicas of a packet to a receiver. If the receiver successfully decodes this packet, it then leverages the SIC technique to cancel its interference in other time slots, thereby increasing the ability to decode more packets and resolving collisions successfully. In IRSA, the number of replicas is derived by

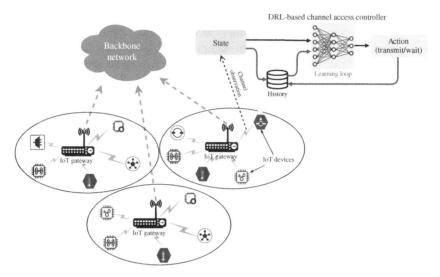

Figure 6.3 DRL-based model for IoT MAC with massive access.

sampling from a probability distribution, which is statically optimized by a differential evolution algorithm. Therefore, IRSA is prone to poor performance in dynamic environments, e.g. changes in network topology and channel condition. In addition, the lack of fully observing the network's state is another challenge when adopting IRSA for WSNs. To address these challenges, the authors propose to leverage the decentralized partial observation Markov decision process (Dec-POMDP) framework [19] to optimize the number of replicas for each sensor to maximize the total system throughput. Unlike the conventional MDP which involves one agent that can fully observe the system state, the Dec-POMDP consists of multiple agents; each can only observe a different part of the system state. In this work, time is divided into frames, and each consists of D equal-size time slots. Each sensing device sends at most one packet per frame. At the beginning of time slot t, an observation of an agent (i.e. a sensing device) is defined as the current number of packets in its buffer. Since agents could not fully observe a system state in Dec-POMDP, they use a belief state that is defined as a finite set of consecutive observations. An action is the number of replicas for a packet sent at time slot t, and an immediate reward for performing an action is the buffer size after sending this packet. The reward for taking an action is a negative number of packets in the transmission buffer. To help agents find their optimal policy, the authors propose a decentralized algorithm based on Q-learning. In particular, at the beginning of a time frame, a utility-based bipartite graph is constructed based on the collision in the previous time frame. In this graph, utility functions are mapped to Q-functions. Then, an algorithm is employed on the graph to compute an optimal joint action (i.e. actions for all agents) of the global Q-function. In addition, the authors adopt the multiple knapsack formulation to reduce the complexity of the algorithm. The simulation results show that the proposed solution achieves a greater throughput than the original IRSA protocol does.

The work in [15] investigated a joint problem of multiple access for a massive number of IoT devices and UAVs' altitude control in a UAV-assisted network, where multiple solar powered-UAVs are deployed as flying BSs. Specifically, IoT devices access the shared uplink (UL) based on an adaptive p-persistent slotted ALOHA to send data to UAVs. In this protocol, time is slotted, and each time slot is further divided into L sub-slots. IoT devices only attempt to access the channel at the beginning of a time slot with a probability of p. To improve access efficiency, the SIC technique is adopted for UAVs to decode received data from IoT devices. During their operation, UAVs may need to ascend to a high altitude to harvest enough energy, and they also must lower their altitudes to improve communication with IoT devices, leading to a conflict problem in UAV altitude control. The channel access probability p of IoT devices and the altitudes of UAVs are managed at every time slot by a central controller connecting to UAVs via a wireless back-haul network. To mitigate the conflict in UAV altitude control and adaptively adjust the channel access probability p of IoT devices under the uncertainty of wireless communications, the authors formulate this problem as a constrained Markov decision process (CMDP). Unlike conventional MDP, the CMDP introduces an additional immediate cost function C to represent constraints. The state is defined as a union of UAVs' altitudes and UAVs' battery levels over h previous subslots, and the characteristics of the two highest received signal-to-interference-plus-noise ratios (SINRs) from associated users at UAVs, i.e. the probabilities of SINRs being higher than a threshold, the mean and variance of SINRs over L communication subslots. By observing the current state at the beginning of a time slot, the central controller takes an action which is a compound of altitude adjustment for all UAVs and the channel access probability for this time slot. After executing an action, the controller receives an immediate reward computed according to the total network capacity and an immediate cost calculated based on the change in battery energy level during this time slot. To find an optimal policy for the central controller, the authors proposed an algorithm based on proximal policy optimization (PPO) and the primal-dual optimization technique. Then, they evaluate their proposed solution via simulations. The results show that the obtained policy can outperform other baselines in terms of the long-term network capacity while guaranteeing the energy sustainability of UAVs.

Instead of leveraging ALOHA-based MAC protocol as in [14, 15], the work in [16] investigated the random access (RA) problem for an IoT network consisting of K IoT devices that are irregularly active to sense and then transmit data to an AP. In particular, the AP needs to allocate N available RBs in the frequency-time grid to active IoT devices to send their sensed data. Note that in general $N \leq K$. In a conventional RA protocol, at the beginning of a time slot, each active device randomly chooses one of M orthogonal sequences to encode its information and then sends it to the AP via a dedicated control channel to indicate its activity status. If a device is classified as active, it will be allocated an RB only if its selected sequence is not chosen by any other devices. An RA procedure is considered as a failure if the number of available RBs is not enough for the active devices, making this approach ineffective since the number of IoT devices is massive. To handle this problem, this work introduces a DRL-aided RA scheme consisting of two stages. In the first stage, the AP allocates N_1 RBs, where $N_1 < N$, according to the DRL algorithm. In the underlying MDP, a state at time slot t represents devices that are active during the t_h previous time slots. Observing this state, the agent (i.e. the AP) determines an action, i.e. a set A_t of IoT devices predicted to be active at this time slot. If the size of A_t is less than or equal to N_1, each device of A_t is given an RB. Otherwise, the AP randomly selects N_1 devices from A_t, and then allocates an RB for each of them. In the second stage in a time slot, the remaining active nodes, i.e. $N_2 = N - N_1$, are allocated based on the conventional RA. At the end of the time slot, the AP takes an action to construct a set S_t consisting of (i) devices that are predicted as active and the AP received data from their allocated RB and (ii) devices detected as active based on conventional RA. Then, a reward for performing this action is K if the agent's prediction is correct and 0 otherwise. In addition, the authors also adopt transfer learning to speed up the learning process. Specifically, the agent is trained in a theoretical model for the activity of the IoT devices and then trained with real-world data. The simulation results show that the proposed approach achieves a greater average packet rate than that of the conventional RA protocol.

Application of DRL-based approaches on IoT devices is still under-explored since most of the works (i.e. [14–16]) focus on theoretical research and evaluate their DRL-based approaches in simulated IoT environment or based on data sets collected in advance. In practice, deploying these solutions directly on IoT devices is infeasible since these devices are highly constrained in resources (i.e. computing, energy, and storage). In addition, wireless channels can be highly dynamic, and operations of radio frequency (RF) components strictly follow timing constraints. However, it may take a DRL algorithm quite sometime, due to the complexity of its DNNs, to converge to the optimal policy. Thus, designing a DRL-based approach to meet the (near) real-time requirement of wireless communications (e.g. in URLLC) is another challenge that needs to be addressed. The study in [17] attempts to bridge these gaps by developing a general-purpose hybrid software/hardware DRL framework, namely, DeepWiERL, which is explicitly designed for embedded IoT devices. In this framework, the training and execution phases of DRL are separated. Recall that a typical DRL algorithm leverages a DNN to learn an optimal policy. After the learning phase, it uses this trained DNN for execution. In DeepWiERL, the execution phase is deployed at the hardware part of an IoT device to meet the real-time constraint, while the training phase is implemented in the software (i.e. edge or cloud) to avoid adding workload to IoT devices with constrained resources. They name the DNN deployed at the IoT hardware as operative deep neural network (ODNN) and the one deployed at edge/cloud as target deep neural network (TDNN). For the underlying MDP, the state is derived from the I/Q samples, which depend on each specific application. An action is defined as a reconfiguration of the wireless protocol stack in the IoT platform. The reward for taking an action is derived according to the objective of a specific scenario (e.g. throughput or latency of data transmission). During the operation of IoT devices, experiences, i.e. a tuple of system sate, action, and reward, are collected to train the TDNN. The ODNN's weights are updated by cloning from those of target DNN at every predefined training step. To guarantee that the TDNN/ODNN model is suitable for the IoT platform to speed up the learning process, the authors proposed the supervised DRL model selection and bootstrap (S-DMSB) method, which includes two techniques, i.e. DNN model selection and transfer learning. To evaluate their proposed framework, they deploy DeepWiERL on a practical software-defined radio platform to address the rate adaption problem. The simulation results show that their solution outperforms the software-based solutions in terms of data rate and energy consumption.

6.2.3 MAC for 5G and B5G Cellular Systems

In earlier cellular generations, e.g. 3G, 4G, the major aim is to provide Mobile Broadband services based on IP connectivity to subscribers. By contrast, 5G aims to support multiple use-cases, each with different network requirements. For example, three classes of use-cases that are covered in 5G are as follows: (i) eMBB to provide high data rate and high traffic volume, (ii) mMTC aiming to support a massive number of low-cost devices with low energy consumption, and (iii) URLLC [20]. In the future, new use-cases with different resource requirements and new technologies will be introduced, leading to a highly complex cellular system where various technologies simultaneously operate. This calls for novel channel access control frameworks for 5G and B5G.

In the literature, several studies attempted to replace the conventional MAC layer in cellular systems with a learnable MAC, e.g. [21, 22]. In [21], the authors considered a single cell system, where L UEs transmit data stored in their buffer to the BS in a shared wireless channel, namely, the UL-SCH channel, as illustrated in Figure 6.4. Data transmissions are controlled by messages exchanged between BS and UEs on uplink and downlink control channels. As such, the UEs and

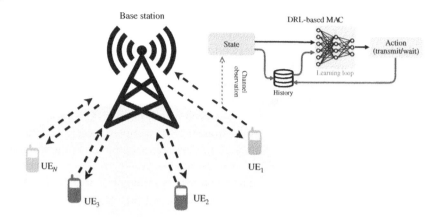

Figure 6.4 DRL-based MAC for 5G.

the BS will learn to leverage exchanged messages on these control channels to maximize the system throughput. To address the dynamicity and uncertainty of wireless networks, they formulate the problem as a Dec-POMDP [19]. Here, the MAC layers of network nodes, i.e. the BS and UEs, are the reinforcement learning (RL) agents. The state of a BS agent is defined as the status of UL-SCH, i.e. (i) idle, (ii) successfully received a packet, and (iii) collision when two or more UEs transmit data in a time slot. The UE state at time slot t is defined as the number of packets waiting in its buffer. At time slot t, the BS agent can decide to do nothing or send one of D available downlink control messages to a UE. At the same time, a UE agent can send one of the U available uplink control messages to the BS, and it can also transmit a packet via the UL-SCH channel. To this end, the UE agents have two types of actions: the communication actions are (i) do nothing and (ii) send one of U uplink control messages; the environment actions are (i) do nothing, (ii) send the oldest packet in its buffer, and (iii) delete the oldest packet in its buffer. The reward r_t for performing an action is determined based on the transmission outcome, i.e. $r_t = \rho$ if the packet is successfully received by the BS and $r_t = -1$, otherwise. To help each agent to learn the optimal policy to maximize the average long-term reward function, which represents system throughput, the authors propose a learning algorithm based on the multi-agent deep deterministic policy gradient (MADDPG) [23] that can simultaneously learn optimal policies for agents. In particular, the MADDPG consists of centralized training and decentralized execution phases. In the centralized training phase, agents share their experiences (e.g. tuples of states, actions, rewards, and next state) during training. In the decentralized execution phase, each agent uses only its observation to make a decision (e.g. transmit/wait). The simulation results show that with collaboration among agents, the proposed approach outperforms the traditional solutions where agents do not communicate with others.

Similar to [21], in [22], the authors focused on wireless channel access policy and signaling functions in the MAC layer of a mobile UE. While a signaling policy determines when to send control message on the control channel, a channel access policy describes when and how to transmit data. The authors' objective is to replace the mobile MAC layer with a learning agent, namely, an MAC learner. In other words, mobile UEs will learn their own MAC protocols instead of being controlled by a BS. They consider that U MAC learners are trained simultaneously in a mobile cell coordinated by one BS with a traditional MAC. All MAC learners share the same uplink (UL) channel and aim to learn a UL access policy that can leverage the MAC signaling available at BS without any prior knowledge about the MAC protocol of BS. This study considers three types of signaling messages at the BS. Particularly, if the BS receives more than one scheduling request (SR) on the UL, it will randomly choose one UE and then send a scheduling grant message to this UE to grant the channel access at this time slot. If the BS successfully receives data from a UE, it then sends back an acknowledgment (ACK) message to this UE. Since each MAC learner's observations about the channel may be different from those of other learners due to their locations, the authors formulate the problem as a POMDP [24]. At time slot t, the observation o_t^u of a MAC learner u is defined as the number of service data units (SDU) in its data queue. The authors define two types of actions that an MAC learner needs to perform simultaneously at a time slot. The first one is the channel access action a_t^u, where an agent determines whether to (i) do nothing, (ii) transmit the oldest SDU in the buffer, or (iii) delete the oldest SDU. The second one is the signaling action n_t^u, where an agent decides whether to send a scheduling request on the UL control channel or not. After performing an action, all MAC learners receive a reward of -1 to push them to finish the learning episode as quickly as possible. To obtain an optimal policy for MAC learners, the authors proposed a Q-learning-based algorithm where they leverage a memory buffer to store a history of observation, actions, and control messages. In addition, two Q-tables are used to approximate Q-values of actions, one for the channel access and another for signaling actions.

Spectrum sharing has been shown to be viable approach to address the radio spectrum scarcity problem. In long-term evolution (LTE) technology, the licensed assisted access LTE (LTE-LAA) [25] has been introduced as an effective solution to improve the LTE system performance by utilizing unlicensed spectrum bands, e.g. 2.4 and 5 GHz bands. Specifically, LTE-LAA requires LTE devices to follow the LBT procedure, where they need to sense an unlicensed channel and occupy it only if it is idle.

In [26], the authors studied the channel access problem in LTE-LAA systems. In particular, they consider the LTE-LAA downlink that consists of multiple small base stations (SBSs) belonging to several operators. This network shares the unlicensed bands with a group of Wi-Fi access points (WAPs). This work aims at optimizing the channel selection and fractional spectrum access (i.e. the time allocated to an SBS on an unlicensed channel during a period) to maximize the total throughput of SBSs while maintaining fairness among Wi-Fi networks and LTE-LAA operators. To avoid under-utilization of the unlicensed spectrum, they adopt a proactive coexistence mechanism that can predict traffic patterns and future off-peak hours. As such, demands can be properly allocated to balance the usage of unlicensed bands as well as decrease the probability of collision among WAPs and SBSs, possibly leading to an increase in total served LTE-LAA traffic load. For a scenario where all SBSs have the same priority, the authors formulate this problem as a noncooperative game, in which the SBSs are modeled as players aiming at maximizing their long-term throughput. Then, to solve this game, they develop an algorithm, namely, the RL-LSTM algorithm, that utilizes a DNN based on the long short-term memory (LSTM) neural network architecture [27] and use the Policy Gradient [28] to train this DNN. Unlike in action-value methods (e.g. Q-learning and DQN), in a policy gradient method, an agent learns a parameterized policy and can choose actions without depending on a value function. A state is defined as a combination of the Wi-Fi traffic load on the unlicensed channels and the SBSs traffic load. Based on a state, RL-LSTM outputs an action that is a sequence of decisions on the future channel selection and the fractional spectrum access over a time window. A reward is calculated as the total throughput achieved over the selected channels. They show that the game with the above RL-LSTM's best responses converge to a mixed-strategy Nash equilibrium. Their simulations are then conducted with real data collected over several days. Compared to a proportionally fair coexistence mechanism and a reactive approach, the proposed solution achieves a higher performance in terms of severed total network traffic and average airtime allocation for LTE-LAA while maintaining the Wi-Fi performance.

Similar to [26], in [29], the authors explored the coexistence problem between LTE and Wi-Fi systems. The conventional LTE/Wi-Fi coexistence methods require exchanging information between these systems via a dedicated channel, which is very challenging in practice. In addition, these methods assume that the information of Wi-Fi systems, e.g. traffic demands and the number of users, is known in advance. To this end, this work aimed at developing an MAC protocol to maximize the LTE network capacity while guaranteeing the fairness of spectrum sharing between LTE and Wi-Fi systems without exchanging information. Specifically, the authors considered a model where an LTE system with saturated traffic utilizes an unlicensed band of a Wi-Fi system whose traffic is unsaturated. To facilitate the coexistence of LTE and Wi-Fi, they developed a duty-cycle spectrum sharing framework, in which time is equally divided into frames with size T_f, and each frame consists of multiple time slots with the same size. At the beginning of a frame, LTE transmissions are allowed to occupy the shared channel for T_L time slots, and then the channel is for Wi-Fi transmissions for the rest of a frame. The value of T_L is crucial for the system's performance. For example, setting T_L to a small value, when Wi-Fi traffic is low, leads to low utilization of the spectrum. In contrast, a large value of T_L, when Wi-Fi traffic is high, can degrade the operation of Wi-Fi. In practice, Wi-Fi load is often unknown in advance, making the selection of the optimal value for T_L very challenging. To address this problem, the authors proposed a DRL-based algorithm that can realize intelligent spectrum sharing while avoiding the adverse impact on the coexisting Wi-Fi system. In particular, they formulated the problem by using an MDP framework, in which an agent (i.e. an LTE BS) needs to determine the value of T_L based on its observation (i.e. the state) at the beginning of each step. Here, a step is defined as the interval T_N during which the number of Wi-Fi devices remains the same, and thus there are multiple frames in a step. A state s_t at step t is defined as a tuple consisting of the channel activity indicator, which is the longest average idle duration denoted by LID_t, the average number of idle slots and busy slots in a frame in the previous step, i.e. $t - 1$. In addition, an action and reward of $t - 1$ are also included in a state s_t. At the end of t, the agent receives a reward r_t if LID_t is larger or equal to a predefined guard interval T_G; otherwise, it will receive a zero reward. To avoid the randomness of Wi-Fi packets transmitted in a frame, Wi-Fi devices only send packets buffered in the previous frames. By doing so, idle slots due to empty buffers only appear at the end of a frame, making the channel activity indicator more accurate. Through the simulations, the proposed solutions achieve an LTE throughput and Wi-Fi packet undelivered ratio that is close to that achievable by the genie-aided exhaustive search (GAES) algorithm [29] which needs to know all the information (e.g. Wi-Fi traffic demand) about the Wi-Fi system.

While the above works [26, 29] used DRL to address unlicensed spectrum sharing between cellular and Wi-Fi networks, the problem of licensed spectrum sharing between the cellular and the device-to-device (D2D) communication systems was explored in [30]. D2D communication can provide a direct high data-rate connection between devices when they are in close proximity, without requiring orchestration from an AP or BS [31]. The authors in [30] considered licensed spectrum sharing for OFDMA-based single-cell downlink communication, which consists of M cellular users (CUEs) and N pairs of D2D users. The D2D users work in the underlay mode (i.e. sharing mode) and use a set of K RBs from licensed spectrum bands. The authors consider that each CUE is allocated an RB which can be also allocated to multiple D2D connections. As such, D2D communications can cause interference to CUEs, and vice versa. This work considers three types of interference: (i) from the BS to D2D users, (ii) from D2D communications to CUEs, and (iii) between D2D connections when they share the same spectrum band.

Each D2D connection needs to optimally select an RB to maximize its throughput while guaranteeing that the SINRs for the CUEs are higher than the required threshold value. Due to the dynamicity of wireless environments, different D2D connections are likely experience different channel conditions, the authors leveraged a partially observable Markov game, which is a multiagent extension of the MDP, to formulate the D2D channel access problem. In this framework, at each time slot t, an agent (corresponding to a D2D transmitter) observes its current surrounding environments (i.e. agent's state s_t) to take an action a_t (i.e. selecting an RB among available RBs) according to its current channel access policy. An agent's state consists of (i) the instant channel information of this D2D link, (ii) cellular link information, (iii) the interference of the previous time slot, and (iv) the RB selected in the previous time slot. After taking an action, if the SINR of a CUE (that operates on the selected RB) is higher than a threshold, an agent receives a reward r_t that equals to the throughput of D2D connection at this time slot. Otherwise, it receives a negative reward. To find the optimal policy for all agents, the authors proposed a framework based on the Actor-Critic method [28], namely, the multi-agent actor-critic (MAAC) method. In MAAC, an agent maintains two DNNs, i.e. an actor for selecting an action and a critic for estimating a state (or a value of executing an action at a state). During the training phase, the critic updates itself according to the feedback from environment (i.e. the new state and reward), and then the actor is updated according to critic's changes. Since training the DNNs requires high-computing resources, the training phase of all agents takes place at the BS. Then, the actors' weights are cloned to the actor of the corresponding agent. Therefore, the agents needs to upload their experiences, i.e. a tuple $\langle s_t, a_t, r_t, s_t \rangle$, to the BS. To improve the stability of the learning process, experiences of all agents are used to train each agent's critic. Observing that the nearby users should be allocated different RBs, while users that are sufficiently apart can share the RBs, the authors proposed the neighbor-agent actor critic (NAAC) method to reduce the computational complexity. In particular, instead of using experiences from all agents, an agent in the NAAC only uses its neighbors' experiences to train the critic and the actor. Simulation results show that the proposed approaches, i.e. MAAC and NAAC, converge faster and achieve a higher sum-rate for the D2D links compared to those achievable with other baseline methods, i.e. DQN, Actor-Critic, and Q-learning, while guaranteeing a low interference at the CUEs. Furthermore, while NAAC converges faster than MAAC, MAAC is more suitable for high-reliability communications since it offers the lowest outage probability for the D2D links.

One important application of 5G and B5G systems will be to facilitate intelligent transportation systems (ITS). In [32], DRL was used to address the channel access control problem for the Internet of Vehicles (IoV). The author considered a long roadway segment, where there are several entry/exit points for vehicles. Along this road segment, multiple Internet of Things gateways (IoT-GWs) powered by batteries are deployed to collect information from sensors (which are placed at fixed locations) and vehicles under their communication ranges via a single hop. In particular, when a vehicle enters the coverage of an IoT-GW, it will exchange its downlink service request and its exit point. If sensors or vehicles sense hazardous situations, they will raise a safety message to notify their connected IoT-GWs. Time is slotted, and at the beginning of a time slot, IoT-GWs forward all information from sensors and vehicles under their coverage to a central server via a dedicated channel. Based on collected information and messages from the central server, an IoT-GW grants a link access to a device (i.e. a sensor or vehicle) to (i) upload its safety message or (ii) continue downloading requested data. If the IoT-GW decides to receive data from the device, it then forwards the safety message to the central server, which will broadcast this message to all other IoT-GWs. The energy consumption of receiving the safety message of an IoT-GW is much smaller than that of the device's requested data transmission. This study aims to find the optimal decisions for all IoT-GWs at each time slot to maintain balanced energy among IoT-GWs, while guaranteeing that requests of vehicles are fully served before they leave the road. To address the uncertainty and dynamics of the vehicle arrival and departure processes and the randomness of the hazardous event arrival, this work formulates the problem as an MDP, in which each IoT-GW server acts as an agent. The state of an IoT-GW is defined as a combination of its remaining energy (the instantaneous battery level), the number of vehicles under its coverage, the remaining content request size, the waiting time of safety message in the buffer of vehicles and sensors, and the distances to its connected vehicles. An action of an IoT-GW is to determine a device (i.e. sensor or vehicle) that will upload its safety message or continue downloading its requested content. An immediate reward function is defined as a negative sum of power consumption for the selected action, waiting time of vehicles that still do not receive any service, total delay of completed service request, a penalty due to the departure of a vehicle that cannot complete downloading its service, and a penalty if any IoT-GW is out of battery. The simulation results show that the proposed solution achieves a greater completed request percentage, a lower mean request delay, and a higher total network lifetime compared with other baseline schedulers.

DRL has also found its applications in different scenarios in vehicle networks, particularly in dealing with the mobility of the vehicles. For example, incorporating the mmWave into ITS is particularly challenging due to the high mobility of vehicles and the inherent sensitivity of mmWave beams to dynamic blockages. In such a case, DRL can be recruited at the BS to learn the fading or blockage profile (e.g. caused by buildings or even bus or larger vehicles) to support the handover and beam association decision-making process. To that end, one can formulate the problem under a semi-Markov decision framework to capture the dynamics and uncertainty of the environment, e.g. fading, blockages, velocity. However, in most cases, we may end up with an MDP with large action and state space that impact the convergence rate to the optimal policy. To speed up the convergence process, one can recruit advanced neural network architectures like double Q-learning, deep-double Q-learning, deep dueling. The key idea of the deep dueling is using two streams of fully connected hidden layers to concurrently train the learning process of the Q-learning algorithm, thereby improving the training process and achieving an outstanding performance for the system. The application of DRL to dealing with the dynamics and uncertainty caused by the high mobility of vehicles can be also explored at the PHY and NET layers.

As an example, the authors in [33] considered a beam association/handover problem for an ITS whose services are provided via 5G. Specifically, each vehicle maintains one LTE connection with an eNodeB to exchange control messages and one mmWave connection with a mmWave base station (mmBS) to obtain ITS services. Each mmBS has several beams to cover different zones in a road. Suppose a vehicle enters a blockage zone or leaves its current beam. In that case, it needs to connect to another beam belonging to the currently connected mmBS or another mmBS to avoid service disruption. The beam association/handover is orchestrated by the eNodeB that controls mmBS via backhaul links. To address the high dynamics and uncertainty of mmWave communications due to the high propagation attenuation, severe sensitivity to blockages, and selective directivity, the authors formulate the problem as a semi-MDP, where each vehicle acts as an agent. The state of a vehicle is defined by a tuple of its received signal strength indicator (RSSI), current beam, speed, and direction. Whenever a vehicle leaves the current zone, the eNodeB takes action, i.e. whether to associate the vehicle with a new beam or not. The reward is defined as the amount of data that the vehicle received from the mmBS during the time it stays in the current zone. To find the optimal beam association policy for all vehicles, the authors propose a parallel algorithm based on Q-learning to find the optimal beam association policy for all vehicles. In particular, the parallel Q-learning consists of multiple learning processes (each for a vehicle) that simultaneously run on the eNodeB to update the global Q-table jointly. Then, the authors show that since the Q-table's size is small and the parallel Q-learning requires a small communication overhead, the computational complexity of the proposed approach is low. The simulation results show that the proposed approach can achieve a greater data rate and a lower disconnection probability compared with those of baseline methods based on Q-learning and DQN.

6.3 Heterogeneous MAC Protocols

Future wireless networks (FWNs) are expected to integrate various access technologies (e.g. cellular, Wi-Fi, and ZigBee). Each network may have a different MAC protocol, which will make the MAC problem in FWN to be very challenging task. Note that, unlike the spectrum sharing problem in [26, 29] where the secondary users can only access the shared channel when the primary users do not occupy it, all users in FWNs have the same priority in accessing the shared channels. Figure 6.5 demonstrates an example of applying DRL to solve the heterogeneous MAC problem. In particular, an uplink to an AP is shared among devices using different MAC protocols (i.e. ZigBee, IEEE 802.11, DRL-based protocols). Without requiring information of MAC protocol of other devices, the DRL-based protocol aims at finding the optimal channel access policy that maximizes the network performance, e.g. total throughput and fairness among devices. In general, for the underlying MDP, a state can be defined as an observation of the channel. After observing a current state s_t at time t, an agent (i.e. a wireless device) takes an action a_t, which is to transmit data or wait. Then, the agent receives an immediate reward based on how the network performs. Thus, the objective of the agent is to find an optimal policy that maximizes the long-term reward function as follows:

$$\pi^* = \operatorname*{argmax}_{\pi} R(\pi) = \operatorname*{argmax}_{\pi} \left[\lim_{T \to \infty} \frac{1}{T} \sum_{t=1}^{T} \mathbb{E}\big(r_t(s_t, \pi(s_t))\big) \right]. \tag{6.3}$$

Several works investigated on the application of DRL for the heterogeneous MAC problem, in which networks with different MAC protocols simultaneously

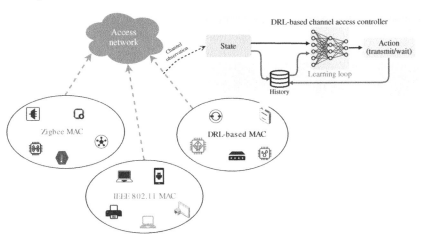

Figure 6.5 Heterogeneous MAC protocols.

operate on shared channels [34–36]. In [34], the authors considered uplink transmissions, in which multiple wireless devices use different MAC protocols (e.g. TDMA and ALOHA) to access a shared channel in a time slotted manner to transmit data to an AP. This study aims at developing two DRL-based MAC (DLMA) protocols that can find an optimal access policy without requiring knowledge about MAC protocols of other devices to maximize the sum-throughput of all the coexisting networks or to achieve α-fairness, respectively. The agents are wireless devices that use DLMA with an objective to maximizing the sum throughput. At time slot t, an agent performs an action $a_t \in \{$ *Transmit, Wait*$\}$ to either transmit or wait, respectively. After all agents take actions, the channel observation is denoted by $o_t \in \{$ *Successful, Collided, Idle*$\}$, where *Successful* indicates that only one agent sends data in t, *Collided* indicates that there is a collision (when more than one agent actually sent data) in t, and *Idle* indicates that no agent sent data in t. Then, the channel-state c_t is defined as $c_t \triangleq (a_{t-1}, c_{t-1})$. The environment-state is defined as a collection of past M channel-states, i.e. $s_t \triangleq [c_{t-M}, \ldots, c_t]$. An immediate reward for taking action a_t is based on o_t, i.e. $a_t = 1$ if $o_t = $ *Successful* and $a_t = 0$ otherwise. To achieve the α-fairness objective, the problem is reformulated by introducing a DLMA gateway node acting as a DRL agent controlling channel access of the DLMA-based network. The state and reward of the DLMA gateway are represented by two vectors. Each of the coordinates in these vectors corresponds to a DLMA device's state and reward, respectively. The action space for the DLMA gateway is the same as that of the DLMA protocol. During a time slot, if the DLMA gateway chooses *Wait*, all DLMA devices will not transmit at this time slot. In contrast, if it chooses *Transmit*, it grants channel access to a DLMA device in a round-robin manner. Simulation results demonstrate that the proposed DLMA protocols can achieve high throughput and maintain proportional fairness without knowing the MAC protocols of other devices.

Extending the work in [34], the study in [36] investigated a heterogeneous MAC protocol in a multichannel scenario. In particular, the authors considered a network consisting of N wireless networks each with a different MAC protocol. The nodes in each network use a dedicated channel for communicating with an AP. The objective of this study is to develop an MAC protocol for an emerging network that does not have any dedicated channel for communicating with the AP. Instead, the devices from the emerging network can leverage the underutilized channels in other networks. Since each network does not have any knowledge of other networks in advance, the authors propose a multi-channel deep-reinforcement learning multiple access (MC-DLMA) for devices in the emerging network, aiming at maximizing the channel utilization of all networks. Specifically, each device in the emerging network acts as an agent in the MDP framework. At a time slot t, an agent decides to wait (i.e. $a_t = 0$) or transmit data on one of N channels, i.e. $a_t = i \in \{1, \ldots, N\}$. Upon receiving a packet, the AP broadcasts an ACK packet to all devices to indicate the transmission status on all channels, i.e. $o_t = [o_t^1, o_t^2, \ldots, o_t^N]$, where $o_t^i \in \{Successful, Collided, Vacant\}$. Then, the authors define the action–observation pair of the agent at time slot t as $z_t = (a_t, o_t)$ and the state s_t at time slot t as M past action–observation pairs, i.e. $s_t = (z_{t-M}, \ldots, z_t)$. The reward r_t for taking action a_t is the total throughput over all channels in time slot t. The simulation results show that MC-DLMA can reach near-optimal performance and achieve a higher sum throughput compared with the baseline method.

Unlike the studies in [34, 36], the work in [35] investigated the spectrum sharing problem under imperfect channels where feedback on transmissions can be corrupted. In particular, the authors considered heterogeneous wireless networks where each wireless network uses a different MAC protocol to access a shared channel in a time-slotted fashion. Each network consists of an AP serving multiple users. The APs are connected via a dedicated collaboration network to exchange control information among them. This work aims at developing an MAC protocol (namely, DLMA) based on DRL for uplink access in a heterogeneous network. This problem is first formulated by using an MDP framework, where each user in the DLMA network is an agent. At time slot t, the agent decides to perform an action a_t, i.e. $a_t = 1$ if it chooses to transmit data and $a_t = 0$, otherwise. To consider imperfect channels, an ACK mechanism is used, where the AP broadcasts an ACK packet consisting of a summary of all DLMA users' transmission status (i.e. success/failure) and throughputs of past K time slots to all users after receiving a packet. A reward r_t for performing action a_t is derived according to an ACK packet from the AP. In particular, $r_t = 0$ if the agent does not transmit data or successfully receives the NACK packet notifying that the transmission is unsuccessful, $r_t = 1$ if the agent successfully receives the ACK packet notifying that the transmission is successful, and $r_t = null$ if the agent does not receive the ACK/NACK packet. In addition, based on the received ACK packet, an observation o_t of the agent at the end of this time slot is defined as B if the channel is used by others, I if the channel is idle, S if the data transmission is successfully, and F if the data transmission is unsuccessful. In a case that agent does not receive the ACK packet, $o_t = null$. A state s_t at time t is defined as a collection of past M channel-state c_t, i.e. $s_t = [c_t, \ldots, c_{t-M}]$. A channel-state c_t is defined as a tuple of action a_{t-1}, observation o_{t-1}, and reward r_{t-1}. At each time slot, an experience (i.e. a tuple of the current state, action, reward, and next state) is stored in a buffer to train a DNN. To address a missing ACK problem due to an imperfect channel, the authors proposed a recovery mechanism in which corrupted experiences containing *null* values are stored in the incomplete-experience buffer. Then, they will be reconstructed based on the next ACK packet that is successfully received. To achieve α-fairness with other networks, the authors proposed a two-stage action selection process. In particular, the agent only sends data if $a_t = 1$ according to the output of its DNN and its throughput is the smallest in the previous time slot. The proposed solution was evaluated in a scenario consisting of one DLMA-based network, one TDMA-based network, and one ALOHA-based network. The simulation results show that DLMA obtains a high throughput while guaranteeing fairness among networks.

6.4 Chapter Summary

In this chapter, we have discussed the application of DRL to address the MAC layer issues in a wireless network. First, we have presented several DRL-based approaches to address the resource allocation problem. Then, the state-of-the-art DRL-based solutions for channel access control in different network technologies, i.e. IEEE 802.11, IoT, and cellular networks, have been described. Finally, we have investigated the application of DRL to address the coexistence of heterogeneous MAC protocols. Thanks to self-learning via interactions with the surrounding environment, DRL-based approaches could mitigate the high dynamics and uncertainty of wireless environments, thus outperforming conventional solutions in the MAC layer. However, the existing DRL-based approaches still pose several limitations. First, the existing approaches considering the application of DRL for the MAC layer often overlook the computational complexities of their proposed solutions, which is a critical factor given the limited resources (e.g. energy and/or computing) of many wireless devices. Second, the convergence of DRL-based solutions are not analyzed theoretically but is usually adopted from the canonical DRL convergence analysis. In other words, these proposed DRL-based solutions may not guarantee convergence to the optimal policy. Future studies need to explore these gaps to enable practical applications of DRL in the MAC layer of wireless networks.

References

1 F. Al-Tam, N. Correia, and J. Rodriguez, "Learn to Schedule (LEASCH): A deep reinforcement learning approach for radio resource scheduling in the 5G MAC layer," *IEEE Access*, vol. 8, pp. 108088–108101, 2020.

2 Y. C. Eldar, A. Goldsmith, D. Gündüz, and H. V. Poor (eds.), *Machine learning and wireless communications*, Cambridge University Press, Aug. 2022.

3 M. Agiwal, A. Roy, and N. Saxena, "Next generation 5G wireless networks: A comprehensive survey," *IEEE Communications Surveys & Tutorials*, vol. 18, no. 3, pp. 1617–1655, 2016.

4 Y. Huang, S. Li, C. Li, Y. T. Hou, and W. Lou, "A deep-reinforcement-learning-based approach to dynamic eMBB/URLLC multiplexing in 5G NR," *IEEE Internet of Things Journal*, vol. 7, no. 7, pp. 6439–6456, 2020.

5 H. Van Hasselt, A. Guez, and D. Silver, "Deep reinforcement learning with double Q-learning," in *Proceedings of the AAAI Conference on Artificial Intelligence*, vol. 30, 2016.

6 T. P. Lillicrap, J. J. Hunt, A. Pritzel, N. Heess, T. Erez, Y. Tassa, D. Silver, and D. Wierstra, "Continuous control with deep reinforcement learning," *arXiv preprint arXiv:1509.02971*, 2015.

7 J. F. Kurose and K. W. Ross, *Computer networking: A top-down approach edition*. Addison-Wesley, 2007.

8 W. Wydmański and S. Szott, "Contention window optimization in IEEE 802.11ax networks with deep reinforcement learning," in *2021 IEEE Wireless Communications and Networking Conference (WCNC)*, pp. 1–6, IEEE, 2021.

9 D. Kotagiri, K. Nihei, and T. Li, "Distributed convolutional deep reinforcement learning based OFDMA MAC for 802.11ax," in *ICC 2021-IEEE International Conference on Communications*, pp. 1–6, IEEE, 2021.

10 V. Mnih, K. Kavukcuoglu, D. Silver, A. A. Rusu, J. Veness, M. G. Bellemare, A. Graves, M. Riedmiller, A. K. Fidjeland, G. Ostrovski, *et al.*, "Human-level control through deep reinforcement learning," *Nature*, vol. 518, no. 7540, pp. 529–533, 2015.

11 E. Khorov, A. Kiryanov, A. Lyakhov, and G. Bianchi, "A tutorial on IEEE 802.11ax high efficiency WLANs," *IEEE Communications Surveys & Tutorials*, vol. 21, no. 1, pp. 197–216, 2018.

12 R. Ali, N. Shahin, Y. B. Zikria, B.-S. Kim, and S. W. Kim, "Deep reinforcement learning paradigm for performance optimization of channel observation–based MAC protocols in dense WLANs," *IEEE Access*, vol. 7, pp. 3500–3511, 2018.

13 H. B. Pasandi and T. Nadeem, "Unboxing MAC protocol design optimization using deep learning," in *2020 IEEE International Conference on Pervasive Computing and Communications Workshops (PerCom Workshops)*, pp. 1–5, IEEE, 2020.

14 E. Nisioti and N. Thomos, "Robust coordinated reinforcement learning for MAC design in sensor networks," *IEEE Journal on Selected Areas in Communications*, vol. 37, no. 10, pp. 2211–2224, 2019.

15 S. Khairy, P. Balaprakash, L. X. Cai, and Y. Cheng, "Constrained deep reinforcement learning for energy sustainable multi-UAV based random access IoT networks with NOMA," *IEEE Journal on Selected Areas in Communications*, vol. 39, no. 4, pp. 1101–1115, 2020.

16 I. Nikoloska and N. Zlatanov, "Deep reinforcement learning-aided random access," *arXiv preprint arXiv:2004.02352*, 2020.

17 F. Restuccia and T. Melodia, "DeepWiERL: Bringing deep reinforcement learning to the internet of self-adaptive things," in *IEEE INFOCOM 2020-IEEE Conference on Computer Communications*, pp. 844–853, IEEE, 2020.

18 S. Sen, N. Santhapuri, R. R. Choudhury, and S. Nelakuditi, "Successive interference cancellation: A back-of-the-envelope perspective," in *Proceedings of the 9th ACM SIGCOMM Workshop on Hot Topics in Networks*, pp. 1–6, 2010.

19 T. Jaakkola, S. Singh, and M. Jordan, "Reinforcement learning algorithm for partially observable Markov decision problems," *Advances in neural information processing systems 7*, 1994.

20 3GPP Tech. Rep. 38.913, "Study on Scenarios and Requirements for Next Generation Access Technologies (Release 14), v14.3.0," 2017.

21 M. P. Mota, A. Valcarce, J.-M. Gorce, and J. Hoydis, "The emergence of wireless MAC protocols with multi-agent reinforcement learning," in *2021 IEEE Globecom Workshops (GC Wkshps)*, pp. 1–6, IEEE, 2021.

22 A. Valcarce and J. Hoydis, "Toward joint learning of optimal MAC signaling and wireless channel access," *IEEE Transactions on Cognitive Communications and Networking*, vol. 7, no. 4, pp. 1233–1243, 2021.

23 R. Lowe, Y. I. Wu, A. Tamar, J. Harb, O. Pieter Abbeel, and I. Mordatch, "Multi-agent actor-critic for mixed cooperative-competitive environments," *Advances in neural information processing systems 30*, 2017.

24 L. P. Kaelbling, M. L. Littman, and A. R. Cassandra, "Planning and acting in partially observable stochastic domains," *Artificial Intelligence*, vol. 101, no. 1–2, pp. 99–134, 1998.

25 X. Wang, S. Mao, and M. X. Gong, "A survey of LTE Wi-Fi coexistence in unlicensed bands," *GetMobile: Mobile Computing and Communications*, vol. 20, no. 3, pp. 17–23, 2017.

26 U. Challita, L. Dong, and W. Saad, "Proactive resource management for LTE in unlicensed spectrum: A deep learning perspective," *IEEE Transactions on Wireless Communications*, vol. 17, no. 7, pp. 4674–4689, 2018.

27 S. Hochreiter and J. Schmidhuber, "Long short-term memory," *Neural Computation*, vol. 9, no. 8, pp. 1735–1780, 1997.

28 R. S. Sutton and A. G. Barto, *Reinforcement learning: An introduction*. MIT Press, 2018.

29 J. Tan, L. Zhang, Y.-C. Liang, and D. Niyato, "Intelligent sharing for LTE and WiFi systems in unlicensed bands: A deep reinforcement learning approach," *IEEE Transactions on Communications*, vol. 68, no. 5, pp. 2793–2808, 2020.

30 Z. Li and C. Guo, "Multi-agent deep reinforcement learning based spectrum allocation for D2D underlay communications," *IEEE Transactions on Vehicular Technology*, vol. 69, no. 2, pp. 1828–1840, 2019.

31 "Samsung 5G vision white paper." Available at: https://www.samsung.com/global/business/networks/insights/white-papers/5g-core-vision/.

32 R. F. Atallah, C. M. Assi, and M. J. Khabbaz, "Scheduling the operation of a connected vehicular network using deep reinforcement learning," *IEEE Transactions on Intelligent Transportation Systems*, vol. 20, no. 5, pp. 1669–1682, 2018.

33 N. Van Huynh, D. N. Nguyen, D. T. Hoang, and E. Dutkiewicz, "Optimal beam association for high mobility mmWave vehicular networks: Lightweight parallel reinforcement learning approach," *IEEE Transactions on Communications*, vol. 69, no. 9, pp. 5948–5961, 2021.

34 Y. Yu, T. Wang, and S. C. Liew, "Deep-reinforcement learning multiple access for heterogeneous wireless networks," *IEEE Journal on Selected Areas in Communications*, vol. 37, no. 6, pp. 1277–1290, 2019.

35 Y. Yu, S. C. Liew, and T. Wang, "Multi-agent deep reinforcement learning multiple access for heterogeneous wireless networks with imperfect channels," *IEEE Transactions on Mobile Computing*, vol. 21, no. 10, pp. 3718–3730, pp. 2021.

36 X. Ye, Y. Yu, and L. Fu, "Multi-channel opportunistic access for heterogeneous networks based on deep reinforcement learning," *IEEE Transactions on Wireless Communications*, vol. 21, no. 2, pp. 794–807, 2021.

7

DRL at the Network Layer

7.1 Traffic Routing

The work in [1] is one of the first attempts to apply deep reinforcement learning (DRL) for optimizing network routing operations. In the proposed DRL system, the environment state is defined as the traffic matrix measuring bandwidth requests between all possible pairs of sources and destinations. The agent interacts with the environment by explicitly providing a routing path for each source–destination pair and receives the reward as the average end-to-end delay. The optimal routing policy, i.e. the policy that minimizes network delays, is solved by leveraging the actor–critic method combined with a deterministic policy gradient algorithm and a stochastic exploration scheme. In the experiment, the DRL agent is trained with traffic matrices of multiple intensity levels generated by using a gravity model [2]. The OMNet+ discrete event simulator [3] is used to analyze the effectiveness of the new approach. The results indicate that the DRL agent is trained genuinely, as the average network delay decreases with training time. It also shows that the average performance of the DRL solution surpasses 75% of 100,000 randomly generated routing configurations. Moreover, the proposed approach offers significant benefits compared to conventional heuristic and optimization techniques as it can rapidly produce near-optimal routing decisions without requiring a model of the underlying system.

Unlike [1], the authors in [4] proposed a DRL-based control framework to deal with traffic engineering (TE) problems in general communication networks. TE is the operation of monitoring and analyzing network status to manage the flow of traffic to attain optimal performance effectively. In particular, the authors in [4] aim to develop a traffic rate allocation strategy, given the demands of ongoing end-to-end communication sessions. A communication session includes a pair of source-destination nodes and a collection of candidate paths that would carry the traffic load. A TE problem solution describes how each communication session's traffic load is split among its candidate paths. This refers to a continuous control

Deep Reinforcement Learning for Wireless Communications and Networking:
Theory, Applications, and Implementation, First Edition.
Dinh Thai Hoang, Nguyen Van Huynh, Diep N. Nguyen, Ekram Hossain, and Dusit Niyato.
© 2023 The Institute of Electrical and Electronics Engineers, Inc. Published 2023 by John Wiley & Sons, Inc.

problem, which can be addressed using the actor-critic approach by Deep Deterministic Policy Gradient (DDPG) algorithm. However, the authors in [4] showed that this conventional method does not handle the targeted TE problem well, as it misses a clear exploration strategy and neglects the importance of transition samples in the replay buffer as a residue to sampling.

To address this problem, the authors in [4] designed a practical DRL-based traffic control framework named DRL-TE, which combines two ideas for improving DDPG, particularly for TE problems, i.e. TE-aware exploration and actor-critic-based prioritized experience replay. TE-aware exploration algorithm randomly selects new action by adding random noise to a base TE solution such as a load balance solution or Network Utility Maximization (NUM) based solution [5]. The idea of using prioritized experience replay, as suggested in deep Q-network (DQN)-based DRL [6], is also applied to improve the performance of the proposed actor-critic framework. Specifically, transitions in the experience replay are assigned different priorities regarding both the critic network's temporal difference (TD) training error and the actor network's Q gradient. Based on the given priority value, the transition is randomly sampled in each epoch, so transitions with higher errors are selected more frequently.

In the proposed DRL model, the system state includes each communication session's delay and throughput measurements. The action defines the ratios of traffic load split among all candidate paths for each communication session. The reward is the total utility of all communication sessions in the network, where the utility function considers both throughput and delay. The DRL-TE framework is tested on the network slicing (NS)-3 simulator using both conventional network topologies and topologies randomly produced by the BRITE generator [7]. Compared to other solutions, such as DDPG and NUM-based solutions, the proposed method significantly reduces the end-to-end delay and enhances the network utility.

Despite the DRL approach having many attractive features, previous DRL-based TE or routing frameworks also face two major issues: hard to converge and low robustness [8]. To address those problems, the authors in [8] propose a Scalable and Intelligent NETwork (SINET) control architecture that combines the DRL framework with pinning control theory. The architecture of SINET is shown in Figure 7.1. In particular, SINET heuristically selects a subset of network nodes, called driver nodes, to perform control policy, thus reducing the state-space complexity of the system. The set of driver nodes is chosen during the offline phase of the framework, where the proportion of driver nodes in the network is derived by use of pinning control theory. In the following online training phase, a DRL framework is employed where the state is described by a matrix showing network performance, each element in the matrix represents the throughput of a traffic flow monitored on a port of a driver node. The agent action defines the

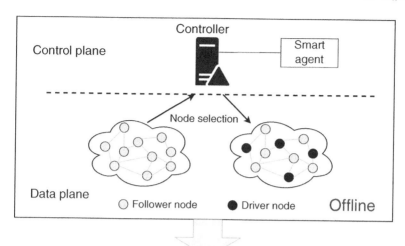

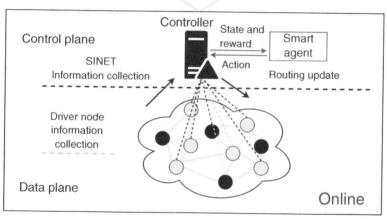

Figure 7.1 An illustration of the SINET architecture. Adapted from Sun et al. [8].

link weight of the driver nodes, one per each monitored port. The reward can be defined as a network performance measurement, such as network utility, network throughput, or transmission delay. Simulation results show that the proposed SINET outperforms other competitors including the DRL-TE and DDPG scheme in terms of flow completion time, especially in the case of large topologies. SINET also offers superior training performance due to smaller state space and better robustness to topology changes.

Considering the benefits of network function virtualization (NFV) technology, the authors in [9] proposed a virtual network function (VNF) placement and traffic routing algorithm regarding real-time network changes based on DDPG, called A-DDPG. The algorithm aims at maximizing a network utility function, which reveals the revenue-cost trade-off associated with the service providers (SPs). It was shown in [9] that the problem can be modeled as an MDP, where the system state is represented by a vector of the available link and node resources through a predefined time period T. An action is a tuple that indicates VNF placement and traffic routing, and the reward is defined as the value of the utility function. By leveraging the concept of attention mechanism, a conventional technique in image processing, A-DDPG will turn its focus to necessary nodes to adjust the behavior of the agent, thus improving training performance. This is done by appending an attention layer before the first fully connected layer of the conventional actor-critic network. Through the experiments, the authors demonstrate that A-DDPG can always achieve the highest training reward compared to the state-of-the-art NFVdeep method, as well as other conventional reinforcement learning (RL) algorithms such as Q-learning and DDPG thanks to the attention mechanism.

The work in [10] considered the NFV technology along with mobile edge computing (MEC) concepts for adapting to the dynamic requirements in IoT networks. Here, the authors studied the problem of planning the placement and routing paths for a sequence of VNFs to minimize the end-to-end delays in a dynamic NFV/MEC IoT network. This problem is termed service function chain (SFC) dynamic orchestration problem (SFC-DOP). The authors then modeled the problem as a Markov stochastic process that can be addressed by the DRL-based method. In the SFC-DOP model, a state consists of the flow rate status of each SFC and the current VNF settlement. An action is a vector of continuous variables that indicate the traffic flow rate on every virtual link, and the reward is defined as an inverse function of the total service delay, which is the sum of the processing delay of VNF on the selected nodes and the transmission delays on the links. An SFC orchestration framework was proposed in [10] based on the DDPG algorithm which includes two main components: the edge-cloud environment and the learning agent, as illustrated in Figure 7.2. The simulation results show that the new proposal achieves significantly lower end-to-end service delay compared with previous schemes including the asynchronous advantage actor-critic (A3C) algorithm, DQN algorithm, and genetic algorithm.

7.2 Network Slicing

7.2.1 Network Slicing-Based Architecture

Network slicing is a key innovating technology in 5G networks, which is also anticipated to play an important role in the following network generation. In the network-slicing concept, multiple separate and independent logical

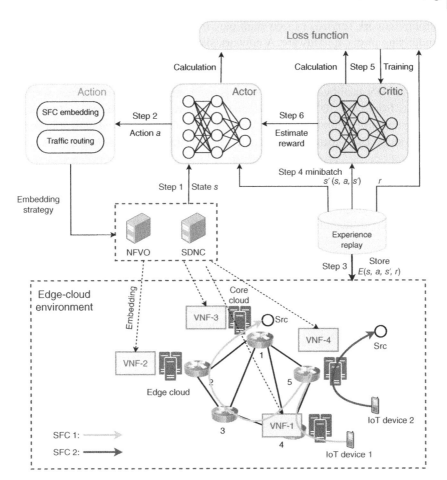

Figure 7.2 An illustration of the SFC orchestration scheme based on DDPG. NFVO; NFV-orchestration, SDNC; SDN-controller. Adapted from Liu et al. [10].

networks, or slices, can coexist on the same physical network infrastructure. Software-defined networking (SDN) and NFV are two principal technologies for enabling network slicing [11]. To improve network performance, SDN uses a centralized controller, which is commonly located in the core cloud, to manage the delivery of traffic flow and orchestrate the allocation of network resources. In an SDN-enabled network slicing system, the SDN controller yields an abstract pool of resources and control logic for creating new slices as in a traditional client-server model. As a result, SDN makes it easier to predefine slice blueprints and to set up slice instances on-demand using the corresponding services and resources. On the other hand, NFV technology leverages the virtualization technique by letting network services run on multiple virtual machines independently from underlying physical hardware. A network slice in NFV can contain one or more virtualized network functions, whereas a network service can be viewed as a part of a network slice [12].

While SDN facilitates network management, network slicing gathers several benefits from both SDN and NFV. First, this lowers the capital expenditures of setting up and running the network by allowing different virtual network operators to share the same physical underlying infrastructure. Additionally, network slicing offers the opportunity to design tailored slices for diverse service types with varied QoS needs, thus can establish service differentiation and ensuring service level agreements (SLAs) for each type of service. Third, network slicing promotes the adaptability and flexibility of network management since slices can be generated on-demand, updated, or deleted as necessary.

7.2.2 Applications of DRL in Network Slicing

Recently, DRL has emerged as a key technique to realize network slicing. In [13], the authors addressed the dynamic resource scheduling problem in radio access network (RAN) slicing while preserving the independence among slices. In the proposed strategy, referred to as the intelligent resource scheduling strategy (iRSS), the authors used both the downlink (DL) and RL approaches in order to solve the resource allocation problem in both large and small time scales. Specifically, the DL technique is used to predict the traffic load coming in the next time window, while the RL method can learn and quickly adapt to the dynamics of the environment. Moreover, the significance between those two approaches can be dynamically adjusted to maintain the accuracy of the proposed collaborative learning framework throughout the entire training process. For the traffic volume prediction task on a large time scale, a long-short-term memory (LSTM) architecture is adopted and trained from the historical traffic records collected by one or some distributed data collector units. For the online resource scheduling task on a small time scale, the authors modeled the decision process as a continuous-time MDP and applied the A3C method to achieve an optimal control policy.

In [14], the authors considered a multi-tenant scenario, where multiple SPs share the same physical infrastructure with an RAN-only slicing, as illustrated in Figure 7.3. To meet the service requirements of multiple subscribed mobile users (MUs), each SP must compete with other SPs for slice resource access, including computation and communication resources. This problem, termed RAN-only cross-slice resource orchestration for multitenant, is then modeled as a stochastic noncooperative game among the SPs. Each player in the game corresponds to an SP who observes the dynamic of the global network to find a control policy on channel bidding, computation offloading and packet scheduling so that its

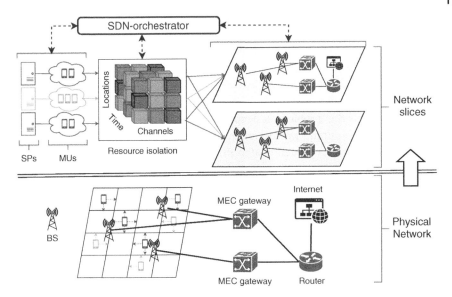

Figure 7.3 An illustration of the proposed radio access network-only slicing architecture. Adapted from Chen et al. [14].

own predicted long-term reward is maximized. However, the computation of an optimal policy for an SP requires knowledge about global network dynamics including the local status of other SPs, which are impractical in a realistic network environment. To overcome this problem, the authors used an abstract stochastic game to model the interactions between the SPs. That is, for each SP, the global network state is conjectured as a local abstract state that is independent of the behavior of other SPs, thus enabling the estimation of an expected future payoff. Next, the MDP model of each SP is linearly decomposed to reduce the complexity of the decision-making process. Based on the formulated MDP, the optimal abstract policy for each SP can be learned by using a double DQN with the experience replay technique. Numerical results show that the proposed DRL-based approach can successfully learn the optimal control policy and it can provide a higher average utility performance of MUs as well as a lower number of packet drops, in comparison with three baseline schemes.

Unlike [14], the authors in [15] proposed a resource management framework that enables the network provider to jointly allocate different kinds of resources such as computing, storage, and radio to multiple slice requests in a real-time fashion. The main focus is handling the heterogeneity, uncertainty, and dynamic nature of slice requests. By modeling the real-time arrival of requests as a semi-Markov decision process (SMDP), the optimal resource allocation policy can be obtained by using the conventional RL approach, such as a Q-learning algorithm. In the proposed SMDP model, the system state is a vector representing the number of running slices in different classes. To maximize the long-term reward, the network provider may choose to accept or refuse new arriving slice requests. However, the authors show that when simultaneously considering combinatorial resources, the Q-learning algorithm faces the curse-of-dimensionality problem as the state/action spaces become large. To address this problem, a deep dueling network architecture is adopted to accelerate the training process of the resource slicing framework. The core concept behind the deep dueling algorithm is to simultaneously train the value and advantage functions using two streams of fully connected hidden layers, as illustrated in Figure 7.4. Experiment results show that, by leveraging the deep dueling architecture, the proposed framework can produce up to 40% higher long-term average rewards in comparison to existing network slicing techniques, as well as a thousand times faster in the training process.

Similarly, the work in [16] considered the implementation of RAN slicing in fog computing enabled 5G networks. A specific network scenario is investigated where two types of network slice instances is jointly orchestrated, namely, the hotspot and the vehicle-to-infrastructure (V2I) slice instance. The authors then formulate the corresponding RAN slicing as a nonlinear integer optimization problem of cooperatively deciding the mode selection and content caching, with respect to the hotspot UEs' bit rate requirements and V2I UEs' transmission delay requirements. To deal with the dynamic wireless channel and uncertain content popularity distribution, an effective solution based on the DRL approach is suggested. In this DRL-based model, the system state consists of the channel state of each user equipment and the cache status of each fog access point. The action of the DRL agent, i.e. the central cloud, is to decide which content should be cached and what transmission modes should be used. The system reward is described as a function that measures how well the optimization target and restrictions were met. Simulation results demonstrate the effectiveness of the new proposal as it can significantly outperform the baselines and is resilient to time-varying channels under the unknown distribution of content popularity.

Unlike the previous works, the authors in [17] investigated the integration of network slicing and device-to-device (D2D) communication to enhance network performance in vehicular networks. This is a challenging task since vehicle-to-vehicle (V2V) communication often requires lower latency, higher throughput, as well as massive connectivity. One idea to address this issue is to aggregate the D2D resources to create a shared resource pool and utilize a network-wide resource allocator to distribute-free resources to slices that are in need to guarantee user demand. More specifically, the considered system architecture is made up of five components: cellular base station (BS), resources, vehicles, slices, and slicing controller, as depicted in Figure 7.5. User scheduling and

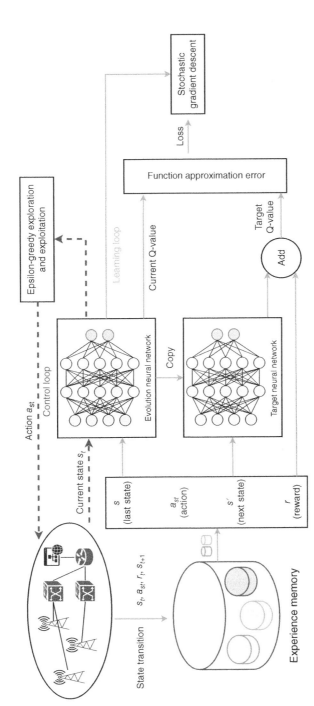

Figure 7.4 An illustration of the deep dueling network architecture used to allocate resources in network slicing.

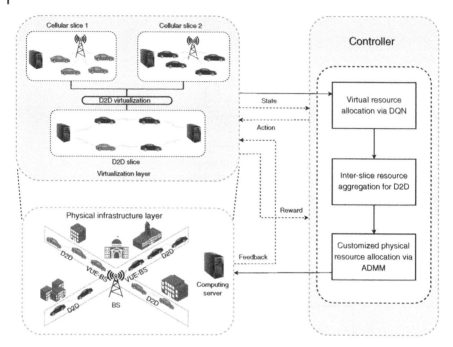

Figure 7.5 An illustration of the proposed resource slicing and customization framework for a vehicular network. VUE; vehicular user, ADMM; alternating direction method of multipliers (ADMM). Adapted from Sun et al. [17].

resource distribution are handled by the BS. The slicing controller, which plays the role of a DRL agent, monitors the status of all slices and channel conditions to perform slice scheduling, resource allocation, and select transmission mode. The proposed slicing framework is adopted a three-level architecture. In the first level, the DRL agent performs autonomous resource slicing by creating/dividing slices and dynamically assigning resources to slices. This DRL agent aims to ensure that the cellular communication performance will not drop below a minimum threshold while also maximizing the sum rate for interslice D2D links. In the next level, the D2D resources are aggregated into a common resource pool for D2D-based V2V communications. Finally, by using the alternating direction method of multipliers, customized physical resource allocation is carried out, whereby resources are assigned to each user according to the slice requirement. Extensive simulations showed that, when compared to previous solutions, the proposed slicing framework can effectively enhance resource usage, slices satisfaction, and throughput gains.

A practical device association mechanism for RAN slicing was proposed in [18]. Inherently, device association involves network slice selection, BS association,

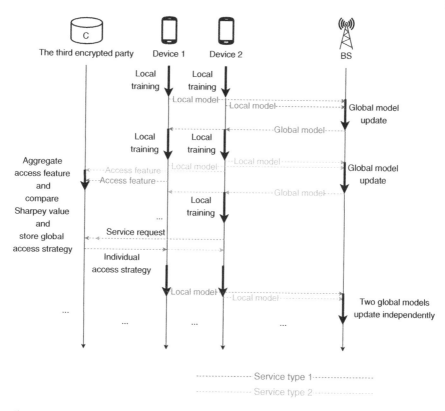

Figure 7.6 An illustration of the proposed hybrid federated deep reinforcement learning process. Liu et al. [18].

and related resource allocation problems. These problems should be considered simultaneously to increase resource utilization while ensuring the quality of service. In [18], the authors combined DRL algorithms with federated learning (FL) framework into a decentralized device association scheme, named hybrid federated deep reinforcement learning (HFDL), that aims at reducing the handoff cost and enhancing network throughput. The architecture of hybrid federated deep reinforcement learning (HDRL) is illustrated in Figure 7.6, which comprises a DRL model running on each device and two layers of model aggregation, namely horizontal weights aggregation (hDRL) and vertical access feature aggregation (vDRL). The DRL model adopts the double deep Q-network (DDQN) algorithm, where the system state includes the currently selected NS/BS of the host device and the bandwidth utilization. The agent action is to specify the amount of bandwidth allocated to the device at a time slot, and the reward is a function indicating the trade-off between communication efficiency and communication cost.

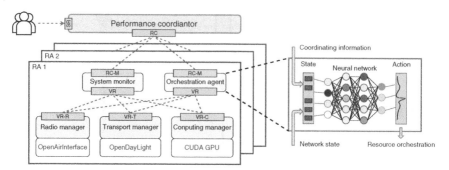

Figure 7.7 An illustration of the distributed resource orchestration system for wireless edge computing networks. CUDA; compute unified device architecture, RC; resource coordination, SDNC; SDN-controller. SR; slice request, VR; virtual resource. Adapted from Liu et al. [19].

The DRL models in multiple devices are trained and aggregated by both hDRL and vDRL to guarantee data security and privacy. In particular, hDRL aggregates the DRL parameters from devices running the same type of service, while vDRL aggregates the local access features from diverse services by using Shapley values to evaluate the importance of features. The proposed HDRL scheme was shown to outperform other state-of-the-art solutions in terms of communication efficiency and network throughput.

In [19], a distributed resource orchestration system for wireless edge computing networks called EdgeSlice was introduced to dynamically perform end-to-end network slicing. EdgeSlice leverages the decentralized DRL approach by utilizing several dispersed orchestration agents and a central performance coordinator as shown in Figure 7.7. In order to maintain the SLA of network slices, the performance coordinator monitors the resource orchestration policies in all orchestration agents. Additionally, there are multiple orchestration agents that understand the resource requirements of the slices and manage the resources appropriately to maximize performance while maintaining network capacity. Each orchestration agent is a DRL agent that monitors the network state including network slice queue length and coordinating information passed from the central performance coordinator. The DRL agent then decides on the number of resources allocated to network slices in the current time interval and receives a reward as a sum of the total system performance and a penalty if the constraints are violated. The performance of the EdgeSlice system was evaluated through both network simulation and an experiential prototype system. The authors showed that EdgeSlice can significantly gain the system performance when compared to other resource orchestration systems.

Exploiting the power of a generative adversarial network (GAN) in approximating data distributions, the authors in [20] proposed a generative adversarial network-powered double-deep Q network (GAN-DDQN) that can effectively learn the action-value distribution in RAN slicing scenario. In GAN-DDQN, the system state describes the number of packets that arrives in each slice, the action is to specify the bandwidth allocated for each slice, and the reward is a system utility. The authors notice that GAN-DDQN can mitigate the negative effects of obtrusive randomness and noise incorporated into the received SLA, satisfaction ratio, and spectrum efficiency. The authors also introduced a reward-clipping strategy to improve GAN-training DDQN's stability. In order to decouple the state-value distribution and the action advantage function from the action-value distribution and overcome the inefficiency of GAN-DDQN in calculating gradient, the dueling structure, which is inherited from the Dueling DQN, is applied to the generator to form an improved model called Dueling GAN-DDQN. Simulations show that GAN-DDQN and Dueling GAN-DDQN perform significantly better than the traditional DQN technique.

Thus far, the existing DRL-based proposals for RAN slicing can only effectively handle pure continuous stochastic optimization problems (e.g. power allocation) or pure integer optimization problems (e.g. slice configuration, radio resource allocation). In [21], the authors aimed at developing a novel DRL framework by integrating conventional DRL solutions for both discrete action space (i.e. deep Q-network) and continuous action space (i.e. Deep Deterministic Policy Gradient) with convex optimization tools to reduce the action space. In particular, by noting that the network condition is influenced by both long-term dynamics (such as traffic demands) and short-term dynamics (such as wireless channel variation), the authors adopted a two-level control strategy for intelligent RAN slicing (Figure 7.8). First, the upper-level controller intends to ensure the QoS requirements of slices as well as the performance isolation between multiple slices. Here, a DQN agent is employed to learn the optimal control policy by other observing the upper-level state that includes the average packet arrival rate, the average packet delay, and the average packet drop rate of UEs. The DQN agent then adjusts the slice configuration by changing the guaranteed and maximum bit rate of active UEs. On the other hand, the lower-level controller will allocate physical resources and transmit power to current UEs. At this level, a modified DDPG framework is applied to optimize the long-term utility of all network slices, where the system state reflects the queue length and channel status of each UE. Additionally, a new strategy to reduce the action space (and hence the computational complexity) is suggested by leveraging the convex optimization techniques. According to the simulation results, the proposed intelligent RAN slicing scheme outperforms the baseline iRSS scheme in terms of QoS and convergence.

The implementation of network slicing in LoRa-based industrial Internet of Things (IIoT) networks was investigated in [22]. To handle heterogeneous QoS

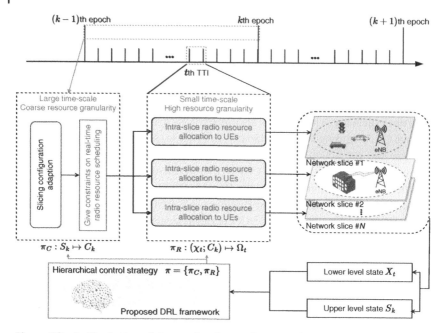

Figure 7.8 An illustration of the two-level control strategy for intelligent RAN slicing. TTI; transmission time interval. Adapted from Mei et al. [21].

requirements of multiservices, a novel deep federated RL framework for network slicing was proposed to dynamically manage and allocate network resources including transmission power (TP) and spreading factor (SF). In the proposed framework (Figure 7.9), multiple DQL agents, each one located in a network gateway (GW), interact with the shared environment to learn the local TP and SF control policy that maximizes the slices' QoS revenues. A local DQL agent observes the state of virtual IoT slices including slices' members and slices' energy efficiency and then performs resource allocation in terms of TP and SF. Besides, there is a central orchestrator employing a global deep federated learning (DFL) model that aggregates the locally trained models from the agents to provide optimal action for maximizing slice revenues. The numerical results demonstrate that the new FL-based framework can rapidly and dynamically serve the QoS requirements of the slices and has superior performance compared to conventional centralized slicing schemes.

The work in [23] highlighted the potential of applying uplink (UL) and DL decoupled RAN framework for cellular V2X networks to improve load balancing, throughput, and reduce energy consumption. The authors propose a two-tier solution for UL/DL decoupled RAN slicing where the DRL approach, namely, the soft actor-critic (SAC) algorithm, is adopted at the first tier to efficiently perform

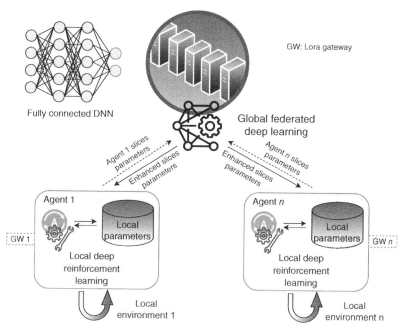

GW: Lora gateway

Fully connected DNN

Global federated deep learning

Agent 1 slices parameters

Enhanced slices parameters

Agent *n* slices parameters

Enhanced slices parameters

Agent 1

Local parameters

GW 1

Local deep reinforcement learning

Local environment 1

Agent *n*

Local parameters

GW *n*

Local deep reinforcement learning

Local environment n

Figure 7.9 An illustration of the deep federated RL framework for network slicing. Adapted from messaoud et al. [22].

bandwidth allocation to different BSs (Figure 7.10). In the associated MDP model, the system state is denoted by a tuple of the bandwidth allocated at the last time slot, the total bandwidth requirement in the current time slot, and the reward from the previous time slot. The DRL agent controls the slicing ratio that determines the bandwidth allocated to each BS and receives reward as a function of the network utility and the QoS of relay-assisted cellular vehicle-to-vehicle (RAC-V2V) communication. At the second tier of the proposed framework, the spectrum bandwidth will be strictly allocated to each vehicular user. Here, the QoS metric for RAC-V2V is converted into an absolute-value optimization problem, which can be resolved by a proposed alternative slicing ratio search (ASRS) algorithm. Numerical results show that the proposed UL/DL-decoupled RAN slicing framework can significantly improve load balancing, reduce C-V2X transmit power, and enhance network throughput.

A new radio and power resources allocation for RAN, named, EE-DRL-RA was proposed in [24], which shares the same idea as the previous iRSS scheme [13], where DRL and DL are used, respectively, to handle the large and small time-scale decision-making problems for resource allocation. However, the EE-DRL-RA scheme improves on the large time-scale prediction accuracy by replacing the

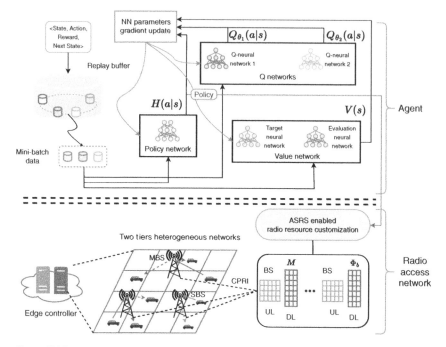

Figure 7.10 An illustration of uplink and downlink decoupled RAN framework for cellular V2X networks. MBS; macro base station, NN; neural network, SBS; small base station. Adapted from Yu et al. [23].

conventional LSTM model in iRSS with its extended version, namely, the stacked bidirectional long-short-term memory (SBiLSTM). By learning the input sequence (both forward and backward) and concatenating both knowledge, the SBiLSTM model can produce more precise predictions with a faster convergence speed. Additionally, the authors formulated the problem of energy-efficient power allocation (EE-PA) as a nonconvex optimization model and solved it using an effective iterative algorithm based on the gradient descent method to select the optimal TP and resource blocks for users requesting rate-based slices. Simulation results show that the EE-DRL-RA scheme outperforms the original iRSS scheme and also the iRSS enhanced with EE-PA, in terms of convergence speed, energy efficiency, the number of served users, and the level of inter-slice isolation.

In [25], the authors proposed to use a graph attention network (GAT) to exploit the collaboration within BSs in the context of a dense cellular network. GAT is known as a powerful tool to process graph-structured data, such as the distribution of BSs in cellular networks. The authors integrate GAT with the widely used DRL methods to produce an intelligent real-time resource allocation scheme for network slicing, where the temporal and spatial dynamics of service demands can be captured and processed by GAT in the state preprocessing stage. Next, a DRL method is adopted to learn the optimal interslice resource management policy. The research problem is modeled by an MDP, where the system state consists of the GAT-processed observation data from the agents, i.e. the BSs. Each agent performs bandwidth allocation for each NS and earns a reward as a specified function that demonstrates whether the lowest satisfaction ratio (SSR) requirement is met. The authors then select two typical types of DRL algorithms, namely, the value-based DRL (DQN and its variants) and the hybrid policy/value-based DRL (advantage actor-critic [A2C]), to evaluate the general capacity of GAT in enhancing the performance of DRL algorithms. Simulation results show that employing GAT on top of these DRL algorithms for state pre-processing, can improve BS collaboration, and provide an efficient policy for RAN slicing.

7.3 Network Intrusion Detection

In conventional intrusion detection scenarios, security engineers need to monitor and analyze audit data such as network traffic flow, application log files, and user behaviors to detect and quickly respond to malicious activities and security threats. However, when the network scale expands, the volume of audit data quickly increases, putting a tremendous workload on the system that cannot be manually handled. This challenging issue can be addressed by installing an intrusion detection system (IDS) on endpoint computers or network equipment, which is a software or hardware platform that monitors audit data to identify and notify the administrator of any suspicious or malicious activity in real-time. Such an automated system for intrusion detection and prevention would be able to act appropriately when recognizing early signs of intrusion attempts to mitigate the consequences of malicious actions.

Generally, IDSs can be divided into two categories: host-based intrusion detection system (HIDS) and network-based intrusion detection system (NIDS), depending on the forms of audit data monitored and processed by the system. An HIDS, which often operates on a single specific operating system, uses some typical kinds of data such as computer settings, application traces, and service log files to identify anomalous activities. Implementing host-based solutions necessitates familiarity with the host operating systems and settings since they typically lack cross-platform compatibility. On the other hand, network-based systems are more portable and are designed to monitor traffic across specified network segments, regardless of the destination operating system. Thus, it is simpler to implement network-based systems as they can significantly reduce deployment costs, while also providing more monitoring capabilities than HIDSs. A network-based IDS, however, would face the scalability problem when handling high-speed networks with heavy traffic loads since it requires packet-level inspection.

In an IDS, signature-based and anomaly-based detection are the two most well-known techniques for identifying abnormal behaviors. A signature-based detection algorithm first constructs a database of known attack patterns and then compares potential attack traits with those in the database. Meanwhile, anomaly based detection algorithms regularly monitor the system state in normal conditions and notify the administrator when an uncommon event occurs, such as an unexpected rise in traffic rate, or the number of network layer packets sent and received per second. Machine learning algorithms, such as unsupervised and supervised classification approaches, can become powerful tools for building adaptive IDSs. However, these techniques, such as logistic regression, support vector machines (SVM), K-nearest neighbors, random forests, and more recently deep neural networks, typically rely on some specified features of already-existing cyberattacks, making them ineffective at handling new or deformed attacks that are not observed in the training data. In this context, RL and DRL techniques can be used as effective methods to deal with dynamic intrusion and provide appropriate action in response to an attack for both HIDS and NIDS.

7.3.1 Host-Based IDS

In [26], the authors proposed a novel IDS based on reinforcement learning that can handle complex intrusion behaviors. To distinguish between normal and abnormal system call traces, a Markov reward process model is employed, where the system states are characterized as short sequences of system calls extracted from a single process trace. Also, each trace is given a reward value such that the reward for detecting a normal trace is −1, while the reward for detecting an abnormal trace is +1. Rewards for all intermediate transitions between states are set at 0. It can be observed that the value function of a state will provide a likelihood estimate of the normality or abnormality of the underlying trace. Thus, the intrusion detection problem is translated into a Markov chain state value prediction task, which can be efficiently tackled by the linear TD RL algorithm. The TD learning method updates the state value function using the differences between subsequent approximations rather than the differences between real values and approximated ones. After the offline training process, the value functions of the states are computed using the trained value prediction model. Then this state value function can be used in the online detection phase to assess if the trace is normal or abnormal by comparing the state value with a predetermined or optimized threshold. Experimental results demonstrate that the proposed RL-based IDS can not only achieve 100% detection precision and zero false alarm rate but also reduce computing costs, in comparison to the Hidden

Markov Models, SVM, and other machine learning techniques. However, the proposed approach, which is based on the linear TD algorithm, has a drawback when the sequence of anomaly observation is strongly nonlinear. To address this issue, the authors in [27] and [28] proposed a kernel-based reinforcement learning (KBRL) technique employing least-squares temporal difference (LSTD) estimator for training RL agent in the Markov chain state value prediction task. Since the kernel LSTD method can estimate anomaly probabilities with accuracy, it helps IDS perform better at detection, notably when handling complex multistage attacks.

7.3.2 Network-Based IDS

The disadvantages of intrusion detection techniques based on signature and anomaly were discussed by Deokar and Hazarnis in [29]. On the one hand, because signature detection relies on a database of acknowledged attack patterns, it is unable to identify novel attack types, even with an attack that slightly differs from the existing attack signatures. Anomaly detection, on the other hand, has a high incidence of false alarms and might not be able to identify intrusive behaviors that mimic regular activity. The authors then proposed an IDS that combines the strengths of both signature-based and anomaly based detection through the usage of log files in order to successfully detect seen and unseen attacks. These objectives are accomplished using the ideas of log correlation, reinforcement learning, and association rule learning. In the proposed RL framework, the learning agent will receive a reward if the log file specified in a rule includes abnormalities or any signs of an attack, otherwise, the agent will get a penalty. Such reward function helps the system to pick log files that are more suitable while looking for signs of an attack.

In [30], the application of multiagent systems and machine learning to protect against DDoS attacks was investigated using distributed rate-limiting techniques. The authors presented a new method for the original multiagent router throttling called coordinated team learning (CTL). The proposed solution incorporates three mechanisms that are task decomposition, hierarchical communication, and team rewards, so that several defense nodes distributed across various places may work together to stop or mitigate the flood. In the basic multiagent reinforcement learning (MARL) design, each agent's state space consists of an aggregate load, which is defined as the total volume of traffic that passed through the router during the monitoring window. In addition, each agent receives the same reward or penalty, which is called a system or global reward function. The basic MARL approach, however, is limited by the "curse of dimensionality" and cannot scale up in large network scenarios. To overcome the aforementioned issue, the authors further proposed the coordinated team learning strategy based on the original multiagent router throttling by applying the divide-and-conquer paradigm. The main idea is to incorporate RL agents into multiple teams and agents in the same team to have the same reward or penalty. Simulation results demonstrate that the proposed method achieves superior robustness and flexibility compared to those of the baseline and state-of-the-art approaches, namely, the baseline throttling technique and the additive-increase/multiplicative-decrease (AIMD) throttling algorithms. The scalability of the new MARL algorithm was illustrated in an experiment consisting of 100 learning agents.

Intrusion detection and defense mechanism against DoS attack in wireless sensor networks (WSNs) were developed by Shamshirband et al. [31] based on game theory and a fuzzy extension of the Q-learning method to determine the optimal player policies. Here, the problem can be seen as a three-player game with a base station, an attacker, and sink nodes. The base station and sink nodes collaborate to develop the best strategy for detecting and defending against application-layer DDoS attacks. A fuzzy Q-learning algorithm is used for updating the training parameters. The architecture of the proposed game-based IDS includes two phases, namely, detection and defense. When an attacker launches a DDoS attack targeting a victim node in the WSN by sending an excessive amount of flooding packets above a predetermined threshold, the triple-player game begins, in which two defensive players, i.e. the sink nodes and base station, maintain a two-level defense against a single attacker player. In particular, the sink node applies the Fuzzy Q-learning (FQL) algorithm to assess the level of access of the attacker to the contents, while the base station analyzes the attack records using the FQL algorithm to respond against the attack. The efficiency of the proposal is investigated using the low-energy adaptive clustering hierarchy (LEACH) protocol and compared with other soft computing solutions. The results demonstrate the effectiveness of the proposed method in terms of detection accuracy, energy usage, and network lifetime.

In [32], Caminero et al. presented a novel adversarial reinforcement learning approach for intrusion detection by incorporating the environment's behavior into the learning process of a conventional RL algorithm. In the proposed RL framework for the IDS problem, the system state is defined as a network traffic sample, the agent's action is to predict the type of intrusion and the reward is associated with the classification accuracy. Initially, the authors employed a random data sampling strategy, where the simulated environment calculates the next state by randomly extracting the new samples from a pre-recorded data set of samples. However, to effectively handle the case of highly unbalanced data sets, a more sophisticated sampling strategy is developed, which can be considered as an adversarial strategy. Here, another RL agent, named the environment agent, is introduced along with the classifier agent, which is used to produce new training samples. The objective of this second agent is to reduce the rewards that the classifier achieves, by trying to increase the frequency of incorrect predictions and making the classifier learn the most difficult cases. Such kind of oversampling approach will help to mitigate classification errors for underrepresented classes by preventing training bias. The proposed DRL algorithm improved with adversarial strategy is termed adversarial environment using reinforcement learning (AE-RL). The performance of AE-RL is evaluated by comparing it with other well-known supervised machine learning algorithms such as Multilayer Perceptron, Multinomial Logistic Regression, AdaBoost, Convolutional Neural Network, as well as conventional DRL algorithms including DDQN, D3QN, and A3C. The experimental results show that the proposed algorithm outperforms the competitors in terms of both weighted accuracy and F1 score.

7.4 Chapter Summary

In this chapter, we have discussed the potential applications of DRL in addressing problems at the network layer in wireless networks. There are three big topics covered, including traffic routing, network slicing, and network traffic classification (or network intrusion detection). For each topic, we have reviewed and compared different approaches proposed in the literature. Compared to the conventional optimization techniques, these approaches perform better, especially in dealing with uncertainty in the surrounding environments (e.g. information about attackers). However, we have observed that most of the current works are evaluated through simulations and based on many assumptions such as full observation of the surrounding environment and no system delay in the learning process. The challenges in deploying DRL applications in real-world wireless networks will be discussed in more detail in Chapter 9.

References

1 G. Stampa, M. Arias, D. Sánchez-Charles, V. Muntés-Mulero, and A. Cabellos, "A deep-reinforcement learning approach for software-defined networking routing optimization," *arXiv preprint arXiv:1709.07080*, 2017.

2 M. Roughan, "Simplifying the synthesis of internet traffic matrices," *ACM SIG-COMM Computer Communication Review*, vol. 35, no. 5, pp. 93–96, 2005.

3 A. Varga and R. Hornig, "An overview of the OMNeT++ simulation environment," in *Proceedings of the 1st international conference on Simulation tools and techniques for communications, networks and systems & workshops*, pp. 1–10, 2008.

4 Z. Xu, J. Tang, J. Meng, W. Zhang, Y. Wang, C. H. Liu, and D. Yang, "Experience-driven networking: A deep reinforcement learning based approach," in *IEEE INFOCOM 2018-IEEE conference on computer communications*, pp. 1871–1879, IEEE, 2018.

5 S. H. Low and D. E. Lapsley, "Optimization flow control. i. Basic algorithm and convergence," *IEEE/ACM Transactions on networking*, vol. 7, no. 6, pp. 861–874, 1999.

6 T. Schaul, J. Quan, I. Antonoglou, and D. Silver, "Prioritized experience replay," *arXiv preprint arXiv:1511.05952*, 2015.

7 A. Medina, A. Lakhina, I. Matta, and J. Byers, "BRITE: An approach to universal topology generation," in *MASCOTS 2001, Proceedings Ninth International Symposium on Modeling, Analysis and Simulation of Computer and Telecommunication Systems*, pp. 346–353, IEEE, 2001.

8 P. Sun, J. Lan, Z. Guo, Y. Xu, and Y. Hu, "Improving the scalability of deep reinforcement learning-based routing with control on partial nodes," in *ICASSP 2020-2020 IEEE International Conference on Acoustics, Speech and Signal Processing (ICASSP)*, pp. 3557–3561, IEEE, 2020.

9 N. He, S. Yang, F. Li, S. Trajanovski, F. A. Kuipers, and X. Fu, "A-DDPG: Attention mechanism-based deep reinforcement learning for NFV," in *2021 IEEE/ACM 29th International Symposium on Quality of Service (IWQOS)*, pp. 1–10, IEEE, 2021.

10 Y. Liu, H. Lu, X. Li, Y. Zhang, L. Xi, and D. Zhao, "Dynamic service function chain orchestration for NFV/MEC-enabled IoT networks: A deep reinforcement learning approach," *IEEE Internet of Things Journal*, vol. 8, no. 9, pp. 7450–7465, 2020.

11 X. Shen, J. Gao, W. Wu, K. Lyu, M. Li, W. Zhuang, X. Li, and J. Rao, "Ai-assisted network-slicing based next-generation wireless networks," *IEEE Open Journal of Vehicular Technology*, vol. 1, pp. 45–66, 2020.

12 W. Zhuang, Q. Ye, F. Lyu, N. Cheng, and J. Ren, "SDN/NFV-empowered future IoV with enhanced communication, computing, and caching," *Proceedings of the IEEE*, vol. 108, no. 2, pp. 274–291, 2019.

13 M. Yan, G. Feng, J. Zhou, Y. Sun, and Y.-C. Liang, "Intelligent resource scheduling for 5G radio access network slicing," *IEEE Transactions on Vehicular Technology*, vol. 68, no. 8, pp. 7691–7703, 2019.

14 X. Chen, Z. Zhao, C. Wu, M. Bennis, H. Liu, Y. Ji, and H. Zhang, "Multi-tenant cross-slice resource orchestration: A deep reinforcement learning approach," *IEEE Journal on Selected Areas in Communications*, vol. 37, no. 10, pp. 2377–2392, 2019.

15 N. Van Huynh, D. T. Hoang, D. N. Nguyen, and E. Dutkiewicz, "Optimal and fast real-time resource slicing with deep dueling neural networks," *IEEE*

Journal on Selected Areas in Communications, vol. 37, no. 6, pp. 1455–1470, 2019.

16 H. Xiang, S. Yan, and M. Peng, "A realization of fog-RAN slicing via deep reinforcement learning," *IEEE Transactions on Wireless Communications*, vol. 19, no. 4, pp. 2515–2527, 2020.

17 G. Sun, G. O. Boateng, D. Ayepah-Mensah, G. Liu, and J. Wei, "Autonomous resource slicing for virtualized vehicular networks with D2D communications based on deep reinforcement learning," *IEEE Systems Journal*, vol. 14, no. 4, pp. 4694–4705, 2020.

18 Y.-J. Liu, G. Feng, Y. Sun, S. Qin, and Y.-C. Liang, "Device association for RAN slicing based on hybrid federated deep reinforcement learning," *IEEE Transactions on Vehicular Technology*, vol. 69, no. 12, pp. 15731–15745, 2020.

19 Q. Liu, T. Han, and E. Moges, "EdgeSlice: Slicing wireless edge computing network with decentralized deep reinforcement learning," in *2020 IEEE 40th International Conference on Distributed Computing Systems (ICDCS)*, pp. 234–244, IEEE, 2020.

20 Y. Hua, R. Li, Z. Zhao, X. Chen, and H. Zhang, "GAN-powered deep distributional reinforcement learning for resource management in network slicing," *IEEE Journal on Selected Areas in Communications*, vol. 38, no. 2, pp. 334–349, 2019.

21 J. Mei, X. Wang, K. Zheng, G. Boudreau, A. B. Sediq, and H. Abou-Zeid, "Intelligent radio access network slicing for service provisioning in 6G: A hierarchical deep reinforcement learning approach," *IEEE Transactions on Communications*, vol. 69, no. 9, pp. 6063–6078, 2021.

22 S. Messaoud, A. Bradai, O. B. Ahmed, P. T. A. Quang, M. Atri, and M. S. Hossain, "Deep federated q-learning-based network slicing for industrial IoT," *IEEE Transactions on Industrial Informatics*, vol. 17, no. 8, pp. 5572–5582, 2020.

23 K. Yu, H. Zhou, Z. Tang, X. Shen, and F. Hou, "Deep reinforcement learning-based ran slicing for UL/DL decoupled cellular V2X," *IEEE Transactions on Wireless Communications*, vol. 21, no. 5, pp. 3523–3535, 2021.

24 Y. Azimi, S. Yousefi, H. Kalbkhani, and T. Kunz, "Energy-efficient deep reinforcement learning assisted resource allocation for 5G-RAN slicing," *IEEE Transactions on Vehicular Technology*, vol. 71, no. 1, pp. 856–871, 2021.

25 Y. Shao, R. Li, B. Hu, Y. Wu, Z. Zhao, and H. Zhang, "Graph attention network-based multi-agent reinforcement learning for slicing resource management in dense cellular network," *IEEE Transactions on Vehicular Technology*, vol. 70, no. 10, pp. 10792–10803, 2021.

26 X. Xu and T. Xie, "A reinforcement learning approach for host-based intrusion detection using sequences of system calls," in *Advances in Intelligent Computing. International Conference on Intelligent Computing, Lecture Notes in*

Computer Science (D. S. Huang, X. P. Zhang, and G. B. Huang, eds.), vol. 3644, pp. 995–1003, Springer, 2005.

27 X. Xu and Y. Luo, "A kernel-based reinforcement learning approach to dynamic behavior modeling of intrusion detection," in *Advances in Neural Networks. International Symposium on Neural Networks, Lecture Notes in Computer Science* (D. Liu, S. Fei, Z. G. Hou, H. Zhang, and C. Sun, eds.), vol. 4491, pp. 455–464, Springer, 2007.

28 X. Xu, "Sequential anomaly detection based on temporal-difference learning: Principles, models and case studies," *Applied Soft Computing*, vol. 10, no. 3, pp. 859–867, 2010.

29 B. Deokar and A. Hazarnis, "Intrusion detection system using log files and reinforcement learning," *International Journal of Computer Applications*, vol. 45, no. 19, pp. 28–35, 2012.

30 K. Malialis and D. Kudenko, "Distributed response to network intrusions using multiagent reinforcement learning," *Engineering Applications of Artificial Intelligence*, vol. 41, pp. 270–284, 2015.

31 S. Shamshirband, A. Patel, N. B. Anuar, M. L. M. Kiah, and A. Abraham, "Cooperative game theoretic approach using fuzzy Q-learning for detecting and preventing intrusions in wireless sensor networks," *Engineering Applications of Artificial Intelligence*, vol. 32, pp. 228–241, 2014.

32 G. Caminero, M. Lopez-Martin, and B. Carro, "Adversarial environment reinforcement learning algorithm for intrusion detection," *Computer Networks*, vol. 159, pp. 96–109, 2019.

8

DRL at the Application and Service Layer

8.1 Content Caching

With the ability to significantly reduce the communication burden of content-delivery services, content caching has been a very effective technique to cope with the explosive growth in mobile traffic. Typically, most users often request a few popular contents in a short time, creating massive traffic load demands. Proactive content caching precaches the popular contents at BSs and edge devices close to users, thereby allowing the contents to be served to the users with little delay and alleviating the heavy traffic burden. However, due to the heterogeneity in user demands, network infrastructure, and communication links, the optimal design of caching policy is a challenging task.

8.1.1 QoS-Aware Caching

Content popularity is one of the key factors that need to be considered when designing caching policy. Since the number of contents is often large and their popularity can vary with time, DQL is an effective technique to address the high-dimensional policy design problem. A DQL scheme was proposed in [1] to improve the performance of a caching system consisting of one BS with a fixed cache size. In the considered system, when a request arrives, the BS will decide if the content should be cached or not, and which of the previously stored contents will be removed. The state space of the employed DQL scheme consists of the features of the stored contents and the newly requested contents. For each content, the features include the number of requests in three different time frames, e.g. short, medium, and long terms. The actions include (i) replacing a previously cached content with a new one, and (ii) keeping the cache unchanged. The reward is set to be the long-term cache hit rate. Furthermore, the DDPG method was employed to train the DQL scheme in [1], and the Wolpertinger

Deep Reinforcement Learning for Wireless Communications and Networking:
Theory, Applications, and Implementation, First Edition.
Dinh Thai Hoang, Nguyen Van Huynh, Diep N. Nguyen, Ekram Hossain, and Dusit Niyato.

architecture [2] was utilized to reduce the size of action space and to ensure that the DQL does not miss any optimal policy. The main parts of the Wolpertinger architecture are actor-critic architecture and the K-Nearest Neighbors (K-NN) method. The actor network's objective is to reduce the action space, whereas the critic network can correct the wrong decisions of the actor-network. Both of these networks are implemented using feedforward neural network (FNNs) and updated by the DDPG method. Simulation results show that the proposed approach can outperform the first-in-first-out scheme by up to 25%.

Generally, in dynamic environments, cache contents must be continuously updated according to the dynamic requests from users (which reflect the cache popularity). A deep learning method was used in [3] to optimize the caching policy, i.e. placement or replacement of cached contents. A DNN is employed to train the optimization algorithm in advance, and then the optimization algorithm is used for caching or scheduling to minimize the delay in real time. In [4], a caching scheme was developed to handle the dynamics of user requests in content delivery networks, taking into account the Time-To-Live (TTL) metric, i.e. cache expiration time. In the considered system, there is a cloud database server serving a number of mobile devices. The mobile devices can query and update the same entries in a single database. The server needs to store all cached queries and their expiration times. If one of the cached records is updated, all cached queries become invalid. A long TTL will burden the cache capacity, whereas a short TTL will significantly increase the latency if the database server is physically distant.

To learn the optimal TTL, the authors in [4] employed a continuous DQL. Unlike the DDPG approach in [1], the Normalized Advantage Functions (NAFs) were used for the DQL-based framework in [4]. The main objective of the continuous DQL is to maximize the Q-function while avoiding a computation-intensive optimization at each step of the learning process. Using the NAFs, a single neural network can be employed to output an advantage term and a value function, thereby significantly reducing the computation cost. The encoding of the queries and the query miss rates constitutes the system states. The reward is calculated using the ratio of the number of cached queries over the system's total capacity. As a result, longer TTLs are encouraged when there are few cached entries, whereas shorter TTLs are preferred when the cached entries take up most of the system capacity. To take into account the incomplete measurements of rewards and states during the online phase, the Delayed Experience Injection (DEI) approach is developed. This DEI approach enables the DQL agent to consider incomplete transitions in the absence of measurements. The learning algorithm is evaluated using Yahoo! cloud serving benchmark [5]. Simulation results show that the proposed approach performs much better than a statistical estimator.

8.1.2 Joint Caching and Transmission Control

By learning the contents' popularity [1] and predicting cache expiration time [4], caching policies and content placement can be designed to improve the efficiency of the caching process. In addition to the caching policy, transmission control is also a crucial aspect of caching design, which governs how the content is delivered from the caches to the end-users, especially under the dynamic channel conditions of wireless environments. Due to the mutual interference in wireless networks with many users, transmission control plays an important role in reducing interference and improving the system's power efficiency. In particular, transmission control decides which contents to be transmitted at a given time, as well as transmission parameters such as power, channel allocation, data rate, and precoding. Therefore, the design of caching and transmission control needs to be considered jointly to improve the efficiency of content delivery in wireless networks with many users.

Approaches such as those in [6–9] were developed to address the joint caching and transmission control problem in wireless networks. However, the framework in [6] assumes invariant channel state information, which might be impractical. To address this issue, DQL frameworks were developed to optimize the joint caching and interference management problem in [7–9]. These works study an MIMO system with low backhaul and cache capacities. For transmission control, the precoding design needs the global CSI to be available at each transmitter. Cache status and CSI from each user are collected by a central scheduler, which is also responsible for controlling transmission and optimizing resource allocation. To alleviate the high demand for data transfer, content caching is employed at individual transmitters, thereby allowing more capacity to update and share the CSI in real time. At the central scheduler, the DQL-based technique is used to lower the demands for real-time CSI as well as the computational complexity of resource optimization. With experience replay in training, a DNN is employed to approximate the Q-function of the DQL. Moreover, the parameters of the target Q-network are periodically updated, thereby stabilizing the learning process. The information is then collected and combined to create the system state, which will then be sent to the DQL agent. Based on this state information, an optimal action is decided by the DQL for the current time slot. The actions include choosing active users and allocating resources among them, whereas the system reward is determined by the total throughput of all users. As an extension from [7] and [8], a DQL-based approach was developed in [9]. In this framework, a CNN-based DQN is employed to jointly optimize the caching and transmission control in more practical conditions where the CSI might be imperfect or delayed. Simulation results show that the proposed approach can significantly improve the performance of the considered MIMO compared to that of the baseline scheme [6]. In particular, the proposed DQN scheme can improve the rate by up to 20%.

Besides interference, application-related QoS and user experience are also important performance metrics of wireless systems. By jointly optimizing the transmission rate and cache allocation, a DQL approach was developed in [10] to improve the Quality of Experience (QoE) for Internet of Things (IoT) devices in content-centric wireless networks. In this framework, the caching conditions, including cached contents and service information, and cached contents' transmission rates are the system states. The DQL agent tries to continuously maximize the QoE or minimize the network cost. To improve the effectiveness of the proposed DQL framework, prioritized experience replay (PER) and DDQN are employed. PER frequently repeats the important transitions, thereby allowing the DQN to learn from samples more efficiently. Meanwhile, with two distinct value functions in different neural networks, the DDQN can stabilize the learning process, thereby preventing the DQN from overestimating as the number of actions increases. It is worth noting that since the target network is periodically copied from the estimation network, these two neural networks are partially coupled. Moreover, the discrete simulator ccnSim [11] was employed to model the caching behavior of the cache server in different graph structures. Simulation results show that the proposed DQL framework can achieve a QoE value double that of a baseline method, i.e. the standard penetration test scheme. Furthermore, the proposed DDQN can achieve a much lower computational complexity than that of the standard one.

Qcrucialo is an important metric for evaluating Virtual Reality (VR) services. In [12], a framework using DQL was developed to jointly optimize transmission control and content caching in wireless VR networks. In the considered system, there are UAVs that capture videos of live games. These videos are then transmitted to small-cell BSs via millimeter-wave (mmWave) downlink backhaul links, and finally, the BSs will stream the games to VR users as illustrated in Figure 8.1. Additionally, the BSs also cache the frequently requested contents to serve users. The proposed approach aims to maximize the users' reliability metric, which is determined by the probability that the delay of content transmission satisfies a target delay. To maximize the system performance, optimal decisions regarding the set and format of cached contents, transmission format, and users' association are needed. To find these optimal strategies, the authors combine the Liquid State Machine (LSM) and the Echo State Network (ESN) to develop a DQL for each BS. The network environment information is stored and used by the LSM to optimize the caching and transmission strategies. However, a limitation of conventional LSMs is that the FNNs are used as the output function, which incurs a high computational cost. To address this limitation,

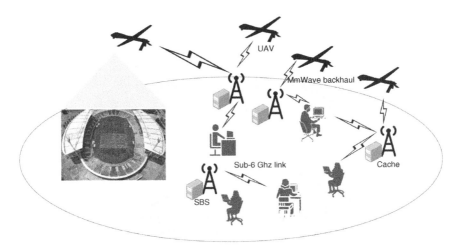

Figure 8.1 An illustration of a VR network assisted by SBSs and UAVs. Adapted from Chen et al. [12]/ IEEE.

the proposed framework replaces the FNNs with an ESN. Using historical information, the ESN can find the relation between the objective, e.g. users' reliability metric, and the caching and transmission strategies. Moreover, the ESN has lower computational demands and a better memory compared to those of the FNNs. Simulation results show that the proposed framework can increase the users' reliability by up to 25.4% compared to that of the baseline Q-learning.

8.1.3 Joint Caching, Networking, and Computation

Typically, there are various communication technologies, such as networked UAVs, vehicular networks, cellular, and device-to-device networks, employed in a HetNet. Such heterogeneity brings forth challenging issues, including interference management, different QoS demands, and resource allocation, that previous approaches for homogeneous networks cannot address efficiently. To address these new challenges, a DQL framework was developed in [13] to minimize the green wireless networks' total energy consumption, taking into account network couplings, computation, and caching. In the considered system, there is a Software-Defined Network (SDN) consisting of multiple virtual networks to serve video requests of mobile users. In particular, a user can request files from nearby small base station (SBSs). The processing of such requests requires computing resources at either the users' devices or the content servers as illustrated in Figure 8.2. To model the wireless channels among the users and the SBSs, the

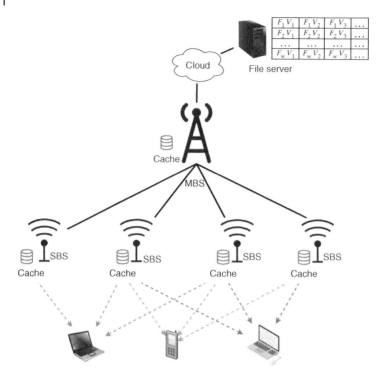

Figure 8.2 An illustration of a video service assisted by computing and caching at the edges. Adapted from [13].

authors adopt the Finite-State Markov Channels (FSMC) model. The states are defined by the channel conditions, the cache capacity, and the computational resource of the users and the content servers. The decisions that the DQL agent at each SBS needs to make include (i) user-to-SBS association, (ii) whether the user or the SBS needs to execute the computational task, and (iii) the scheduling of the data transmissions. Simulation results show that the total energy consumption of the system can be gradually reduced as the learning process carries on. Moreover, compared with DRL schemes that only focus on one of the three tasks, the proposed approach can achieve a significantly lower energy consumption.

Due to its effectiveness, the DQL-based approach in [13] was employed for Vehicular Ad doc NETworks (VANETs) in [14] and [15]. The considered system in [14] consists of multiple BSs, Road Side Units (RSUs), Mobile Edge Computing (MEC), and content servers. A mobile virtual network operator controls all devices in the system. Moreover, the vehicles in the system can request video contents which can be downloaded from the BSs' caches or the remote content servers. A resource allocation problem is formulated, aiming to jointly optimize the caching, computing, and networking decisions for the system. The states are defined by each BS's CSI and computational capability, as well as the cache capacity of each MEC/content server. Using this information, an FNN-based DQN is employed to find the optimal resource allocation for each vehicle. Extended from [14], the authors developed two new schemes in [15] to enhance the performance and stability of the original approach. In particular, to avoid overestimation of the Q values, the authors use a DDQN to decouple the action from the Q-values, and thus the learning speed and stability can be significantly improved. Moreover, since the rewards can sometimes be estimated without taking action, a dueling DQN approach is developed. In the proposed dueling DQN, the state-action Q-value is replaced by two functions. The first function, namely the advantage function, is defined by the relative importance of an action compared to the others. The second function, i.e. the value function, represents the reward. Simulation results show that the proposed approach can improve the total utility by up to 1.6 times compared to that of the static approach.

Similar to [14] and [15], a DQN-based framework was proposed in [16] to jointly optimize communication, computation, and caching for VANETs, taking into account the limited computational resources and storage capacities of the vehicles and the RSUs. The proposed framework is divided into two DQNs for different timescales. The small timescale DQN aims to maximize the reward at every time slot. On the other hand, the large timescale DQN estimates the reward for every certain number of time slots, taking into account the vehicle's mobility. Simulation results show that the proposed framework can save up to 30% of the total system cost compared to that of a random resource allocation scheme.

In [17], a DQL framework was developed for dynamic control of networking, caching, and computation of multiple services in smart cities. In particular, the physical wireless network can be divided into multiple virtual subnetworks using Network Function Virtualization (NFV) [18]. Based on that, a central controller can make decisions regarding network slicing, scheduling, as well as computing, and caching resources allocation. Moreover, DQL can also be utilized to improve the security and efficiency in other smart city applications, such as data exchange and mobile social networks [19, 20].

8.2 Data and Computation Offloading

The limited capacity of tiny IoT devices, e.g. wearable and handheld devices, limits their usage in applications that require high computation resources and storage capacities, e.g. face recognition and interactive online gaming. To circumvent this issue, IoT devices can offload their computational tasks to nearby BSs, APs, MEC servers, or even neighboring mobile users (MUs). For computation-intensive IoT applications, offloading can help to significantly reduce the processing delay, save more battery energy, and improve the system's security. However, determining the amount of computational workload, i.e. offloading rate, and choosing the MEC server for offloading is not an easy task. In particular, if the chosen MEC server does not have sufficient computing resources or the wireless channel between the IoT device and the MEC server is unstable, it might even lead to a worse delay. Therefore, various factors such as user mobility, computational workload, time-varying channel conditions, computing resource capacity, and energy supplies need to be taken into account when designing offloading. Optimization, such as dynamic programming and heuristic, has been developed to find optimal offloading decisions [21]. However, a common limitation of these approaches is that they require a priori knowledge of the users' mobility patterns, which might not be obtainable in practice. To address such uncertainties, DQL can be used. In [22], a DQL-based framework was developed to reduce energy consumption and mobile user costs. In the considered system, mobile users can choose to access either the cellular network or a WLAN, each with a different cost (as illustrated in Figure 8.3(a)). If the data are not successfully sent before a certain deadline, the mobile user has to pay a penalty. The authors propose to model this data offloading decision as an MDP. The system state is defined by the remaining data's size and the mobile user's physical location. The actions of the mobile user include choosing either the cellular network or the WLAN to transmit and the allocation of channel capacities. In the proposed DQL, CNNs are employed to predict the remaining data size of the mobile user. Simulation results show that the DQL-based framework can significantly reduce the energy consumption compared to that of the dynamic programming algorithm, e.g. by up to 500 Joules.

In [23], a DQL scheme was developed for mobile users in an MEC-enabled cellular system, aiming at minimizing power consumption and delay. As shown in Figure 8.3(b), each mobile user has a computing task that needs to be offloaded to the MEC server via a cellular BS. Moreover, each task has different CPU cycles, computational resources, and delay demands. Meanwhile, the MEC server's resources, including computational resources and bandwidth, are limited. To minimize the power consumption and delay taking into account these constraints, a DQL is proposed. The states are defined by the total cost of the system and the idle computing resources of the MEC server. The actions include resource allocation and offloading decisions. Moreover, to reduce the action space, the authors employ a preclassification step to avoid infeasible actions. Simulation results show that the proposed scheme can reduce the total cost of the static allocation strategies by up to 55%.

Unlike [22] and [23] which considered only a single BS, the work in [24] considered multiple BSs, as shown in Figure 8.3(c). As a result, mobile users can choose which BS to transmit their tasks to. On the one hand, the use of multiple

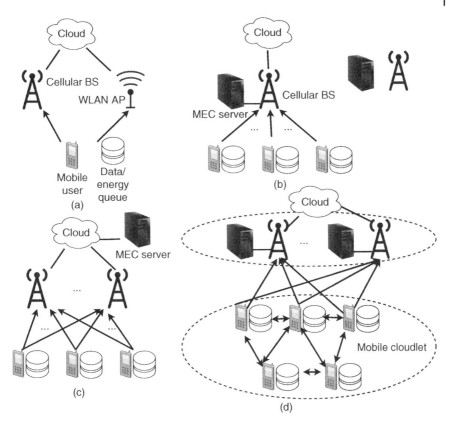

Figure 8.3 Illustrations of data/computation offloading models in mobile networks (a) an MU can choose to access either the cellular network or a WLAN, (b) an MU has a computing task that needs to be offloaded to the MEC server via a cellular BS, (c) an MU can choose one of the BSs to transmit data, and (d) computational tasks can be offloaded to nearby mobile users in an ad-hoc manner.

BSs can help to alleviate heavy traffic and shorten the transmission time. On the other hand, it further increases the complexity of the decision-making process, since the mobile users now have to choose which BS to transmit their tasks to minimize the delay. Under dynamic network conditions, the authors formulate the decision-making processes of mobile users as an MDP. The system states are defined by the channel conditions between each pair of mobile users and BS, the energy consumption, and the task backlogs. The cost function can be determined by the execution and handover delay, as well as the computational task dropping cost. To address this problem, a DDQN-based DQL algorithm was developed in [24] to learn the optimal offloading policy without requiring knowledge of the network's dynamic conditions. Simulation results show that the proposed approach can significantly improve the offloading performance compared with several baseline policies.

Besides additional complexity, the presence of multiple SBSs also has a limitation, i.e. they consume a lot of unnecessary energy in idle time. To address this issue, a DQL-based strategy was developed in [25] to make switching decisions, i.e. turning on and off, for different SBSs, aiming at significantly reducing the energy consumption with minimal effects on the QoS. In particular, a DQL scheme was employed to approximate the value and policy functions. The reward is defined by energy consumption, the costs of switching SBSs, and the decrease in QoS. A DDPG approach is also utilized to speed up the training process. Simulation results show that the proposed scheme can achieve significantly higher computational efficiency with much lower energy consumption compared to those of other baseline methods.

In [26], an RL framework was developed for cloud-based malware detection, where mobile users have the ability to offload their malware detection tasks to the cloud server. In particular, mobile devices with limited energy supply, storage capacity, and computational resources are vulnerable to zero-day attacks [27] as they cannot maintain a huge database for malware detection. By leveraging the cloud server's resources, malware detection performance can be significantly improved. To this end, the authors in [26] modeled the problem as a malware detection game where each player, i.e. mobile user, chooses its offloading rate to the server. In this game, each mobile user employs a DQL scheme to decide its offloading data rate. The system states are defined by the size of application traces and the channel state. The reward is defined by the detection accuracy, which is a concave function over the number of malware samples. To improve the detection speed, the authors employed a CNN to estimate the Q-values of the DQL scheme. Moreover, a hotbooting Q-learning technique is developed to utilize offloading experiences in similar scenarios to improve the initialization phase for Q-learning. Compared to the all-zero initialization strategy, the hotbooting technique can save a significant amount of exploration time and improve learning [28]. Simulation results show that the proposed scheme can improve the detection speed by up to 35.5% while saving more battery compared to those of the baseline methods, including standard Q-learning schemes and hotbooting Q-learning.

In [29], a DQL framework was developed for energy-harvesting IoT networks. The considered system consists of a number of MEC servers, each with different communication and computation capabilities, e.g. BSs and APs, and the IoT devices with different energy storage and energy harvester have to decide their offloading policies. These IoT devices can choose to either offload the computing task or compute locally. This offloading decision is formulated as an MDP. The system states are defined by the predicted amount of harvested energy, the channel capacity, and the battery status. The reward is defined by the energy consumption, overall delay, data sharing gains, and task drop loss in each time slot. A hotbooting Q-learning technique was also employed, similar to [26], to improve the initial learning phase. Moreover, a fast DQL offloading scheme was developed to improve the learning speed of the CNN. Simulation results showed that the proposed approach achieves a performance up to 3.5 times better than that of standard Q-learning. In [30], a two-layered RL algorithm was developed to find optimal task offloading decisions for mobile devices with limited resources, aiming to jointly optimize the processing delay and the machine utilization rate. Specifically, the authors first employed the K-nearest neighbors (KNN) algorithm to cluster the physical machines. Then, the first layer of the two-layered algorithm determines which cluster to offload the task using a DRL scheme. Based on this assignment, the second layer determines a specific machine among the cluster to offload the task. Simulation results show that the proposed approach can converge faster than both the standard Q-learning and DRL. However, the reward gained by the proposed approach is lower than that of the DRL.

Unlike previous works that consider data and computation offloading in cellular systems, the authors in [31] and [32] focused on ad hoc mobile networks. In particular, as shown in Figure 8.3(d), computational tasks in an ad-hoc mobile network can be offloaded to nearby mobile users, i.e. mobile cloudlets. This offloading problem can be formulated as a constrained MDP in [32], in which the states are defined by the cloudlets' available resources, the number of remaining tasks, and the communication links between the cloudlets and mobile users. The main objective is to maximize the total number of tasks processed, constrained by users' QoS which depends on the processing delay and energy consumption. Two schemes, namely, linear programming and Q-learning, were developed to find the optimal offloading decisions. Simulation results show that both proposed schemes can significantly outperform other baseline methods. The authors in [33] considered the same system model to minimize the offloading cost. To this end, a DRL scheme was developed which can take into account uncertainties due to users' and cloudlets' movements. The problem was formulated as an MDP in which the states are defined by the task queues and the distances between the users and the cloudlets. The objective is to maximize the number of tasks processed while minimizing energy consumption. A DQN is then developed to find the optimal offloading decisions. Simulation results confirm the effectiveness of the proposed approach.

Offloading of data and computation is a basic operation of fog computing, where mobile applications are hosted in a container, such as a fog node's virtual machine. Since the users frequently move, the containers must also be migrated or offloaded to other nodes. Moreover, some nodes can be turned off to save energy. In [34], the container migration problem was modeled as a multidimensional MDP, and a DQL is developed to solve this problem. The system states are defined by migration cost, power consumption, and delay. The actions are choosing the containers to be migrated and where to migrate the containers. The authors then propose to classify the nodes based on their level of utilization, and a node can be turned off if it is underutilized. Moreover, to improve learning speed and stability, the authors also employed DDQN and PER. Simulation results show that the proposed approach can achieve lower delay, energy consumption, and migration cost compared to other baseline methods while needing less training time.

8.3 Data Processing and Analytics

In the era of big data, the role of data processing and analytics is significant. However, the increasing complexity and dimensions of data have brought forth many challenges to data processing and analytics, especially for conventional rule-based algorithms [35]. To address these challenges, DRL has emerged to be an effective solution. In this section, we discuss the applications of DRL in many aspects of data processing and analytics, including data partitioning, scheduling, tuning, indexing, and query optimization.

8.3.1 Data Organization

8.3.1.1 Data Partitioning
Data partitioning is an important part of data organization, which can significantly affect the data processing speed and system effectiveness. Finding an optimal partitioning strategy, however, is a challenging task due to a large number of factors such as data characteristics, system implementation, workload, and hardware. In data analytics systems, data are often divided into blocks which will be accessed as a whole by queries. Typically, a query redundantly fetches many blocks, and thus effective block layouts are essential to improving the system performance. In [36], a DRL-based framework was developed to partition data into a tree based on the analytical workload. The tree construction problem is formulated as an MDP with the states defined by a leaf node of the tree. An action is defined by the cutting of the original node to divide it into two child nodes, and the reward is calculated by the number of blocks that can be skipped during a query. Experiment results show that the proposed approach can significantly reduce the run time and error rate compared to other baseline methods.

Besides tree partitioning, horizontal partitioning is also a common approach where data in large tables are distributed across multiple machines depending on certain attributes. The design of this type of partitioning can vary greatly, depending on the predicted run time from experienced database administrators (DBAs) or cost models. To train these models, a lot of data are needed, which is costly and

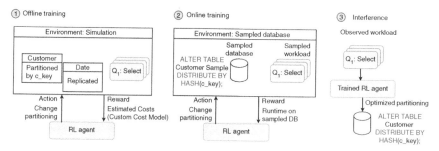

Figure 8.4 An illustration of DRL approach for horizontal data partitioning in a cloud system. Adapted from [37].

time-consuming. To address this issue, in [37], a DRL scheme was proposed to find optimal horizontal partitioning policies in cloud databases. The states in the proposed DRL scheme depend on two factors. First, the database factor is determined by the chosen attribute, the replication of the tables, and table co-partitioning. The workload factor depends on the frequency of queries. The actions include choosing an attribute to partition a table, the replication of a table, and the copartitioning of tables, whereas the reward is minus the workload. To reduce the cost of partitioning during training, the agent is first trained offline using a cost model, and then when used online, a sampling technique is employed to estimate rewards, as illustrated in Figure 8.4. Experimental results show that the proposed approach outperforms classical approaches for several different types of workloads.

Compared to the cases of structured data, partitioning user-defined functions (UDFs) analytics workloads is more challenging, especially on unstructured data due to complex computations and the absence of functional dependency. To address these challenges, an automatic partitioning scheme was proposed in [38], which maps the problem into a graph where partition candidates are the subgraphs. DRL is then employed to find the optimal partition candidates. The state is defined by features such as the most recent run time, key distribution, complexity, frequency, size of co-partition, the execution interval, and hardware configurations, from previous runs. The action is selecting one potential candidate, and the reward is calculated by the difference in throughput between different runs. Historical latency statistics are used to calculate the reward, thereby reducing the running time. Experimental results confirm the proposed approach's effectiveness with different data sizes in different environments.

8.3.1.2 Data Compression

When properly implemented, data compression can save huge storage space, thereby significantly improve the performance of data processing systems. Data

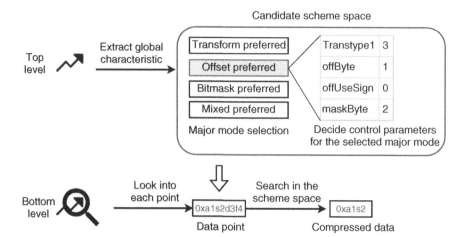

Figure 8.5 Illustration of a two-level framework for data compression. Adapted from [39].

types and patterns are the two main factors that impact the effectiveness of a data compression scheme. For example, since patterns can vary significantly in time-series data, fixed compression schemes are often not effective in such situations. To address this issue, a two-level compression framework was developed in [39], which selects a compression scheme for each point using DRL, as illustrated in Figure 8.5. The states are defined by the time stamp, data header, metrics value, and data segment. The action is to create a setting, whereas the reward is defined by the compression ratio. Experimental results show that the approach can significantly improve the compression ratio, i.e. by up to 50% on average.

8.3.2 Data Scheduling

Scheduling plays a key role in improving resource utilization and maximizing throughput in data processing systems. Typically, job scheduling in such systems depends on a wide variety of factors, including job dependencies, priority, workload size, data location, and hardware configurations. Due to such complexity, general heuristic approaches often do not perform well. To address this issue, a DRL approach was proposed in [40] to find the optimal stage-dependent job schedule. In many data processing systems, jobs might be processed in parallel in numerous stages. To handle the jobs' parallelism and dependencies, a graph neural network (GNN) is first employed to extract the features which are then used to define the states. The actions include determining the next stage to schedule and the degree of parallelism, whereas the reward is set to be the average job completion time. To improve the effectiveness of training in an online manner, the proposed method first finds the schedule for a short sequence of jobs and then gradually increases the number of jobs considered. Experimental results show that the proposed approach can reduce the job completion time by up to 21% compared to other baseline approaches.

In distributed stream data processing, scheduling algorithms need to find the optimal assignment of workers to process streams of continuous data in real-time. Moreover, for each worker, the algorithm also needs to decide the number of threads to use. To find the optimal solution for such complex problems, a DRL algorithm was developed in [41]. The states include the current worker assignment and data tuple arrival rate. The action comprises of assigning threads to machines (workers), whereas the reward is negative value of the average processing time. Since the action space is large, DQN does not perform well. To address this issue, DDPG was employed to train an actor-critic network. Among the output actions of the actor-critic network, the best action is selected. Experimental results show that the proposed approach can reduce the processing time by up to 33.5% and 14% compared to other baseline methods. Besides job assignment, RL-based approaches such as [42–46] were also developed for query scheduling.

8.3.3   Tuning of Data Processing Systems

The configurations of data processing systems have a significant impact on their performances. However, tuning these configurations is challenging because of the very high number of parameters and their inter dependencies. Moreover, factors such as workload and hardware also have an impact on the system performance. Search-based or learning-based methods are often employed to tune these parameters. However, search-based methods often take a lot of time, whereas learning-based methods, e.g. [47], require a lot of high-quality data. To address these challenges, a DRL-based tuning approach was developed in [48]. First, the authors introduced a data cloud tuning system in which the DRL approach is deployed at the cloud server to control all actions of the cloud data center, as illustrated in Figure 8.6. In the proposed DRL scheme, the state is defined by the buffer size and the number of pages read. The action is adjusting knob values, and the reward is defined by throughput and latency. Experimental results show that the proposed approach can outperform [47] under six different workloads on four database settings. However, the query workload is not taken into account.

To address the query workload issue, a DRL scheme was proposed in [49] for tuning the database while taking into account the query information. In the proposed scheme, features such as types, tables, and operations costs are first extracted from the queries. Then, a DNN model is employed to predict the changes of states (defined by the data tuples and the number of transactions).

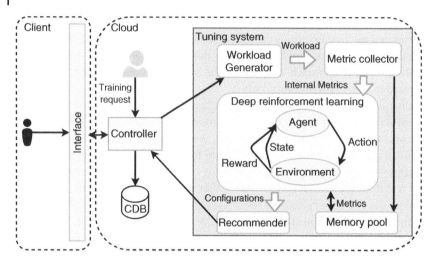

Figure 8.6 Illustration of a data cloud tuning system using DRL. Adapted from [48].

Similar to [48], the action and reward designed in [49] adjust the knob values and system performance (throughput and latency), respectively. Moreover, three levels of tuning granularity were proposed. For query-level tuning, every query is taken into account separately. In contrast, all queries are merged to be used as a single input in workload-level tuning. For cluster-level tuning, a deep learning approach is first employed to classify queries into clusters, and then these clusters can be used as inputs. Experimental results show that the proposed approach outperforms the state-of-the-art tuning methods.

8.3.4 Data Indexing

8.3.4.1 Database Index Selection

The index selection problem, i.e. choosing attributes to create indices, plays a key role in maximizing query performance. In [50], a DRL approach was developed to find an index for a given workload. The states are defined by the selectivity values, the current indices, and the columns in the database. The action comprises of creating an index for a column, and the reward is the gap in performance compared to that of the case with no index. Experimental results show that the proposed approach can perform better than the cases where there is an index for every column. In [51], a DRL approach was developed to find indices for a cluster database, taking into account the query processing and load balancing. Task-specific knowledge was utilized to optimize the action space for the index selection task in [52]. A common limitation of these works is that they only consider single-column indices. To address this issue, a DRL approach was developed in [53] to find indices for single-attribute and multiattribute cases. Additionally, five heuristic rules were developed to reduce the state and action space. However, a limitation of this work is that experiments are only conducted with small and simple datasets.

8.3.4.2 Index Structure Construction

Hierarchical index structures, including the B+ tree and R-tree, can significantly improve the data searching speed. Compared to the single-dimensional indexing mechanisms, it is more complex to optimize the R-tree because of the multipath traversals and bounding box efficiency. Conventionally, heuristics are often utilized to make two decisions while building the tree, i.e. choose which subtree to insert and split the existing node [54]. In [55], a DRL approach was developed to replace the heuristic in R-Tree construction, which consists of two separate MDPs for the two decisions, as illustrated in Figure 8.7. For the subtree MDP, the state is defined by the concatenation of four features, namely, perimeter, occupancy, area, and overlap of each selected child node. The action comprises of inserting a selected node, prioritizing nodes with a high area, and the reward is defined by the performance gap compared with the previous states. For the split MDP, the state is defined by the perimeters and areas of two nodes created by splitting a node with a small area. The action comprises of the selection of the split rule, whereas the reward is similar to that of the subtree MDP. Experimental results show that the proposed approach can improve the performance compared to that of the conventional R-Tree by up to 78.6%.

To improve the speed of nearest neighbors search, graphs can be an effective indexing mechanism [56, 57]. In [58], a DRL approach was developed to optimize the nearest neighbors search's graph, aiming at maximizing the search efficiency. In the proposed approach, the states are defined by the search algorithm and the initial graph. Moreover, the action is edge removal, and the reward is calculated based on the search performance. Experimental results show that the proposed approach can provide improved performance compared to the state-of-the-art graphs. However, the approach can only remove edges from existing graphs, i.e. it cannot add new edges, and thus its performance is limited by the initial graph.

Finding and building new index structures has also gained attention [59]. Taking inspiration from NAS (Neural Architecture Search) [60], an RNN-powered RL neural index search (NIS) framework was developed in [61] to find optimal index structures and related parameters. In particular, the proposed approach can automatically design tree-like index structures layer-by-layer by formulating the index structure as abstract ordered and unordered blocks. Moreover, the proposed approach can also tune all parameters of the building blocks. Experimental results show that the proposed approach can produce an index structure which outperforms the state-of-the-art index.

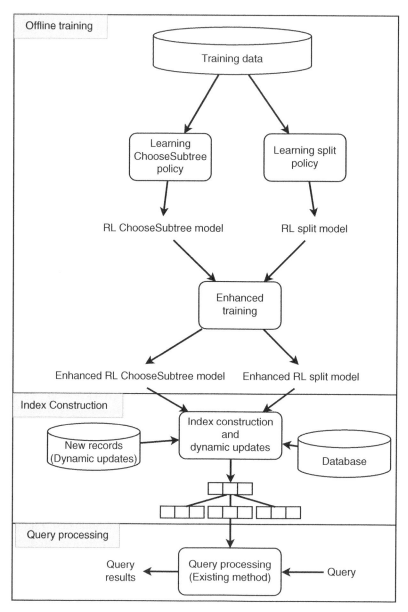

Figure 8.7 The process of proposed RL based R-Tree (RLR) to answer queries. Adapted from [55].

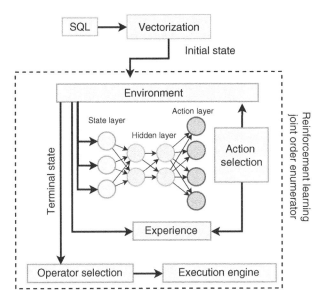

Figure 8.8 An illustration of the proposed DRL framework for join order enumeration. Adapted from [63].

8.3.5 Query Optimization

There can be different ways to execute queries, resulting in substantially different running time. Therefore, optimal query plans can significantly improve the processing time, e.g. from hours to seconds. One of the key factors determining the effectiveness of query plans is the table join orders. Conventionally, heuristics and dynamic programming are often employed to find potential join plans, and cost models are then used to evaluate these plans. However, their performance is limited [62]. To address this issue, a DRL approach was developed in [63] to find optimal join plans. The authors first introduce a new framework, called ReJOIN, as illustrated in Figure 8.8. In this framework, the states are defined by the join predicates and join tree structure. Moreover, the action comprises of combining two subtrees, whereas the reward is calculated using a cost model. Experimental results show that the proposed approach can outperform the PostgreSQL optimizer.

In [64], an extensible featurization approach was proposed for state representation, and in [65], the training efficiency was enhanced by using the DQN algorithm. Moreover, several RL methods such as DQN [65], PPO [66], and DDQN [67] were analyzed to optimize the join order. Additionally, the authors proposed to employ a symmetric matrix instead of a vector for a more expressive state representation. Furthermore, a graph neural network with DRL was developed in [68] for join order selection. In the proposed approach, the state representation is learned by the GNN, resulting in better join tree structure information. Generally, the main difference among these works stems from the state representation, i.e. which information is chosen, and how it is encoded.

Unlike existing approaches that learn from previous experience, the RL approach in [69] optimizes the query execution by learning from the current execution status. In particular, the query execution is divided into small time slots. At the beginning of each time slot, a join order is chosen, and the execution progress is measured. The upper confidence bounds applied to trees (UCT) algorithm in [70] and an adaptive query processing strategy similar to that in Eddies [71] were then adopted. Moreover, a quality metric is introduced, which compares the best query execution with the expected execution. Then, the reward is determined by the progress during the time slot. Additionally, an execution engine is specifically designed for the proposed approach. Experimental results show that the proposed approach performs better than several state-of-the-art approaches, although it also incurs more overheads.

Unlike the previous approaches that aim at replacing the traditional query optimizers, the approach in [72] aimed at improving the traditional query optimizers using RL. Specifically, the proposed approach learns to find the best plans among the candidates produced by traditional query optimizers. Moreover, the query plan trees are transformed into vectors, and then a CNN is employed to find patterns in the tree. Next, the author formulated the plan selection problem as a contextual multiarmed bandit problem. The Thompson sampling technique [73] was then employed to solve the problem. Experimental results show that the proposed approach can outperform a commercial optimizer in terms of running cost and performance.

8.4 Chapter Summary

We have discussed potential applications of DRL to address problems at the application and service layer in wireless networks. Three main topics, i.e. content caching, computation offloading and data processing and analysis, have been discussed in detail. We observe that to tackle these problems, DRL is an effective tool that can deal with the uncertainty and dynamics of requests from users. However, most of the works are evaluated through simulations. Therefore, experimental works would be required to validate the actual benefits of deploying DRL in future wireless networks.

References

1 C. Zhong, M. C. Gursoy, and S. Velipasalar, "A deep reinforcement learning-based framework for content caching," in *2018 52nd Annual Conference on Information Sciences and Systems (CISS)*, pp. 1–6, IEEE, 2018.

2 G. Dulac-Arnold, R. Evans, P. Sunehag, and B. Coppin, "Reinforcement learning in large discrete action spaces. CoRR abs/1512.07679 (2015)," *arXiv preprint arXiv:1512.07679*, 2015.

3 L. Lei, L. You, G. Dai, T. X. Vu, D. Yuan, and S. Chatzinotas, "A deep learning approach for optimizing content delivering in cache-enabled HetNet," in *2017 International Symposium on Wireless Communication Systems (ISWCS)*, pp. 449–453, IEEE, 2017.

4 M. Schaarschmidt, F. Gessert, V. Dalibard, and E. Yoneki, "Learning runtime parameters in computer systems with delayed experience injection," *arXiv preprint arXiv:1610.09903*, 2016.

5 B. F. Cooper, A. Silberstein, E. Tam, R. Ramakrishnan, and R. Sears, "Benchmarking cloud serving systems with YCSB," in *Proceedings of the 1st ACM Symposium on Cloud Computing*, pp. 143–154, 2010.

6 M. Deghel, E. Baştuğ, M. Assaad, and M. Debbah, "On the benefits of edge caching for MIMO interference alignment," in *2015 IEEE 16th International Workshop on Signal Processing Advances in Wireless Communications (SPAWC)*, pp. 655–659, IEEE, 2015.

7 Y. He and S. Hu, "Cache-enabled wireless networks with opportunistic interference alignment," *arXiv preprint arXiv:1706.09024*, 2017.

8 Y. He, C. Liang, F. R. Yu, N. Zhao, and H. Yin, "Optimization of cache-enabled opportunistic interference alignment wireless networks: A big data deep reinforcement learning approach," in *2017 IEEE International Conference on Communications (ICC)*, pp. 1–6, IEEE, 2017.

9 Y. He, Z. Zhang, F. R. Yu, N. Zhao, H. Yin, V. C. Leung, and Y. Zhang, "Deep-reinforcement-learning-based optimization for cache-enabled opportunistic interference alignment wireless networks," *IEEE Transactions on Vehicular Technology*, vol. 66, no. 11, pp. 10433–10445, 2017.

10 X. He, K. Wang, H. Huang, T. Miyazaki, Y. Wang, and S. Guo, "Green resource allocation based on deep reinforcement learning in content-centric IoT," *IEEE Transactions on Emerging Topics in Computing*, vol. 8, no. 3, pp. 781–796, 2018.

11 Q. Wu, Z. Li, and G. Xie, "CodingCache: Multipath-aware CCN cache with network coding," in *Proceedings of the 3rd ACM SIGCOMM Workshop on Information-Centric Networking*, pp. 41–42, 2013.

12 M. Chen, W. Saad, and C. Yin, "Echo-liquid state deep learning for 360° content transmission and caching in wireless VR networks with cellular-connected

UAVs," *IEEE Transactions on Communications*, vol. 67, no. 9, pp. 6386–6400, 2019.

13 Y. He, Z. Zhang, and Y. Zhang, "A big data deep reinforcement learning approach to next generation green wireless networks," in *GLOBECOM 2017-2017 IEEE Global Communications Conference*, pp. 1–6, IEEE, 2017.

14 Y. He, C. Liang, Z. Zhang, F. R. Yu, N. Zhao, H. Yin, and Y. Zhang, "Resource allocation in software-defined and information-centric vehicular networks with mobile edge computing," in *2017 IEEE 86th Vehicular Technology Conference (VTC-Fall)*, pp. 1–5, IEEE, 2017.

15 Y. He, N. Zhao, and H. Yin, "Integrated networking, caching, and computing for connected vehicles: A deep reinforcement learning approach," *IEEE Transactions on Vehicular Technology*, vol. 67, no. 1, pp. 44–55, 2017.

16 R. Q. Hu *et al.*, "Mobility-aware edge caching and computing in vehicle networks: A deep reinforcement learning," *IEEE Transactions on Vehicular Technology*, vol. 67, no. 11, pp. 10190–10203, 2018.

17 Y. He, F. R. Yu, N. Zhao, V. C. Leung, and H. Yin, "Software-defined networks with mobile edge computing and caching for smart cities: A big data deep reinforcement learning approach," *IEEE Communications Magazine*, vol. 55, no. 12, pp. 31–37, 2017.

18 B. Han, V. Gopalakrishnan, L. Ji, and S. Lee, "Network function virtualization: Challenges and opportunities for innovations," *IEEE Communications Magazine*, vol. 53, no. 2, pp. 90–97, 2015.

19 Y. He, F. R. Yu, N. Zhao, and H. Yin, "Secure social networks in 5G systems with mobile edge computing, caching, and device-to-device communications," *IEEE Wireless Communications*, vol. 25, no. 3, pp. 103–109, 2018.

20 Y. He, C. Liang, F. R. Yu, and Z. Han, "Trust-based social networks with computing, caching and communications: A deep reinforcement learning approach," *IEEE Transactions on Network Science and Engineering*, vol. 7, no. 1, pp. 66–79, 2018.

21 C. Zhang, B. Gu, Z. Liu, K. Yamori, and Y. Tanaka, "Cost-and energy-aware multi-flow mobile data offloading using Markov decision process," *IEICE Transactions on Communications*, vol. 101, no. 3, pp. 657–666, 2018.

22 C. Zhang, Z. Liu, B. Gu, K. Yamori, and Y. Tanaka, "A deep reinforcement learning based approach for cost-and energy-aware multi-flow mobile data offloading," *IEICE Transactions on Communications*, vol. E101-B, no. 7, p. 2017CQP0014, 2018.

23 J. Li, H. Gao, T. Lv, and Y. Lu, "Deep reinforcement learning based computation offloading and resource allocation for MEC," in *2018 IEEE Wireless Communications and Networking Conference (WCNC)*, pp. 1–6, IEEE, 2018.

24 X. Chen, H. Zhang, C. Wu, S. Mao, Y. Ji, and M. Bennis, "Optimized computation offloading performance in virtual edge computing systems via deep reinforcement learning," *IEEE Internet of Things Journal*, vol. 6, no. 3, pp. 4005–4018, 2018.

25 J. Ye and Y.-J. A. Zhang, "Drag: Deep reinforcement learning based base station activation in heterogeneous networks," *IEEE Transactions on Mobile Computing*, vol. 19, no. 9, pp. 2076–2087, 2019.

26 L. Xiao, X. Wan, C. Dai, X. Du, X. Chen, and M. Guizani, "Security in mobile edge caching with reinforcement learning," *IEEE Wireless Communications*, vol. 25, no. 3, pp. 116–122, 2018.

27 A. S. Shamili, C. Bauckhage, and T. Alpcan, "Malware detection on mobile devices using distributed machine learning," in *2010 20th International Conference on Pattern Recognition*, pp. 4348–4351, IEEE, 2010.

28 Y. Li, J. Liu, Q. Li, and L. Xiao, "Mobile cloud offloading for malware detections with learning," in *2015 IEEE Conference on Computer Communications Workshops (INFOCOM WKSHPS)*, pp. 197–201, IEEE, 2015.

29 M. Min, L. Xiao, Y. Chen, P. Cheng, D. Wu, and W. Zhuang, "Learning-based computation offloading for IoT devices with energy harvesting," *IEEE Transactions on Vehicular Technology*, vol. 68, no. 2, pp. 1930–1941, 2019.

30 L. Quan, Z. Wang, and F. Ren, "A novel two-layered reinforcement learning for task offloading with tradeoff between physical machine utilization rate and delay," *Future Internet*, vol. 10, no. 7, p. 60, 2018.

31 D. Van Le and C.-K. Tham, "Quality of service aware computation offloading in an ad-hoc mobile cloud," *IEEE Transactions on Vehicular Technology*, vol. 67, no. 9, pp. 8890–8904, 2018.

32 D. Van Le and C.-K. Tham, "A deep reinforcement learning based offloading scheme in ad-hoc mobile clouds," in *IEEE INFOCOM 2018-IEEE Conference on Computer Communications Workshops (INFOCOM WKSHPS)*, pp. 760–765, IEEE, 2018.

33 S. Yu, X. Wang, and R. Langar, "Computation offloading for mobile edge computing: A deep learning approach," in *2017 IEEE 28th Annual International Symposium on Personal, Indoor, and Mobile Radio Communications (PIMRC)*, pp. 1–6, IEEE, 2017.

34 Z. Tang, X. Zhou, F. Zhang, W. Jia, and W. Zhao, "Migration modeling and learning algorithms for containers in fog computing," *IEEE Transactions on Services Computing*, vol. 12, no. 5, pp. 712–725, 2018.

35 Q. Cai, C. Cui, Y. Xiong, W. Wang, Z. Xie, and M. Zhang, "A survey on deep reinforcement learning for data processing and analytics," *arXiv preprint arXiv:2108.04526*, 2021.

36 Z. Yang, B. Chandramouli, C. Wang, J. Gehrke, Y. Li, U. F. Minhas, P.-Å. Larson, D. Kossmann, and R. Acharya, "Qd-tree: Learning data layouts for big data analytics," in *Proceedings of the 2020 ACM SIGMOD International Conference on Management of Data*, pp. 193–208, 2020.

37 B. Hilprecht, C. Binnig, and U. Röhm, "Learning a partitioning advisor for cloud databases," in *Proceedings of the 2020 ACM SIGMOD International Conference on Management of Data*, pp. 143–157, 2020.

38 J. Zou, P. Barhate, A. Das, A. Iyengar, B. Yuan, D. Jankov, and C. Jermaine, "Lachesis: Automated generation of persistent partitionings for big data applications," 2020.

39 X. Yu, Y. Peng, F. Li, S. Wang, X. Shen, H. Mai, and Y. Xie, "Two-level data compression using machine learning in time series database," in *2020 IEEE 36th International Conference on Data Engineering (ICDE)*, pp. 1333–1344, IEEE, 2020.

40 H. Mao, M. Schwarzkopf, S. B. Venkatakrishnan, Z. Meng, and M. Alizadeh, "Learning scheduling algorithms for data processing clusters," in *Proceedings of the ACM Special Interest Group on Data Communication*, pp. 270–288, 2019.

41 T. Li, Z. Xu, J. Tang, and Y. Wang, "Model-free control for distributed stream data processing using deep reinforcement learning," *arXiv preprint arXiv:1803.01016*, 2018.

42 S. Banerjee, S. Jha, Z. Kalbarczyk, and R. Iyer, "Inductive-bias-driven reinforcement learning for efficient schedules in heterogeneous clusters," in *International Conference on Machine Learning*, pp. 629–641, PMLR, 2020.

43 Y. Gao, L. Chen, and B. Li, "Spotlight: Optimizing device placement for training deep neural networks," in *International Conference on Machine Learning*, pp. 1676–1684, PMLR, 2018.

44 T. Kraska, M. Alizadeh, A. Beutel, E. H. Chi, J. Ding, A. Kristo, G. Leclerc, S. Madden, H. Mao, and V. Nathan, "SageDB: A learned database system," 2021.

45 L. Wang, Q. Weng, W. Wang, C. Chen, and B. Li, "Metis: Learning to schedule long-running applications in shared container clusters at scale," in *SC20: International Conference for High Performance Computing, Networking, Storage and Analysis*, pp. 1–17, IEEE, 2020.

46 C. Zhang, R. Marcus, A. Kleiman, and O. Papaemmanouil, "Buffer pool aware query scheduling via deep reinforcement learning," *arXiv preprint arXiv:2007.10568*, 2020.

47 D. Van Aken, A. Pavlo, G. J. Gordon, and B. Zhang, "Automatic database management system tuning through large-scale machine learning," in *Proceedings of the 2017 ACM International Conference on Management of Data*, pp. 1009–1024, 2017.

48 J. Zhang, Y. Liu, K. Zhou, G. Li, Z. Xiao, B. Cheng, J. Xing, Y. Wang, T. Cheng, L. Liu, *et al.*, "An end-to-end automatic cloud database tuning system using deep reinforcement learning," in *Proceedings of the 2019 International Conference on Management of Data*, pp. 415–432, 2019.

49 G. Li, X. Zhou, S. Li, and B. Gao, "QTune: A query-aware database tuning system with deep reinforcement learning," *Proceedings of the VLDB Endowment*, vol. 12, no. 12, pp. 2118–2130, 2019.

50 A. Sharma, F. M. Schuhknecht, and J. Dittrich, "The case for automatic database administration using deep reinforcement learning," *arXiv preprint arXiv:1801.05643*, 2018.

51 Z. Sadri, L. Gruenwald, and E. Lead, "DRLindex: Deep reinforcement learning index advisor for a cluster database," in *Proceedings of the 24th Symposium on International Database Engineering & Applications*, pp. 1–8, 2020.

52 J. Welborn, M. Schaarschmidt, and E. Yoneki, "Learning index selection with structured action spaces," *arXiv preprint arXiv:1909.07440*, 2019.

53 H. Lan, Z. Bao, and Y. Peng, "An index advisor using deep reinforcement learning," in *Proceedings of the 29th ACM International Conference on Information & Knowledge Management*, pp. 2105–2108, 2020.

54 B. C. Ooi, R. Sacks-Davis, and J. Han, "Indexing in spatial databases," *Unpublished/Technical Papers*, 1993.

55 T. Gu, K. Feng, G. Cong, C. Long, Z. Wang, and S. Wang, "The RLR-Tree: A reinforcement learning based R-tree for spatial data," *arXiv preprint arXiv:2103.04541*, 2021.

56 W. Dong, C. Moses, and K. Li, "Efficient k-nearest neighbor graph construction for generic similarity measures," in *Proceedings of the 20th International Conference on World Wide Web*, pp. 577–586, 2011.

57 Y. A. Malkov and D. A. Yashunin, "Efficient and robust approximate nearest neighbor search using hierarchical navigable small world graphs," *IEEE Transactions on Pattern Analysis and Machine Intelligence*, vol. 42, no. 4, pp. 824–836, 2018.

58 D. Baranchuk and A. Babenko, "Towards similarity graphs constructed by deep reinforcement learning," *arXiv preprint arXiv:1911.12122*, 2019.

59 S. Idreos, K. Zoumpatianos, S. Chatterjee, W. Qin, A. Wasay, B. Hentschel, M. Kester, N. Dayan, D. Guo, M. Kang, *et al.*, "Learning data structure alchemy," *Bulletin of the IEEE Computer Society Technical Committee on Data Engineering*, vol. 42, no. 2, 2019.

60 B. Zoph and Q. V. Le, "Neural architecture search with reinforcement learning," *arXiv preprint arXiv:1611.01578*, 2016.

61 S. Wu, X. Yu, X. Feng, F. Li, W. Cao, and G. Chen, "Progressive neural index search for database system," *arXiv preprint arXiv:1912.07001*, 2019.

62 V. Leis, A. Gubichev, A. Mirchev, P. Boncz, A. Kemper, and T. Neumann, "How good are query optimizers, really?" *Proceedings of the VLDB Endowment*, vol. 9, no. 3, pp. 204–215, 2015.

63 R. Marcus and O. Papaemmanouil, "Deep reinforcement learning for join order enumeration," in *Proceedings of the 1st International Workshop on*

Exploiting Artificial Intelligence Techniques for Data Management, pp. 1–4, 2018.

64 S. Krishnan, Z. Yang, K. Goldberg, J. Hellerstein, and I. Stoica, "Learning to optimize join queries with deep reinforcement learning," *arXiv preprint arXiv:1808.03196*, 2018.

65 V. Mnih, K. Kavukcuoglu, D. Silver, A. Graves, I. Antonoglou, D. Wierstra, and M. Riedmiller, "Playing Atari with deep reinforcement learning," *arXiv preprint arXiv:1312.5602*, 2013.

66 J. Schulman, F. Wolski, P. Dhariwal, A. Radford, and O. Klimov, "Proximal policy optimization algorithms," *arXiv preprint arXiv:1707.06347*, 2017.

67 T. P. Lillicrap, J. J. Hunt, A. Pritzel, N. Heess, T. Erez, Y. Tassa, D. Silver, and D. Wierstra, "Continuous control with deep reinforcement learning," *arXiv preprint arXiv:1509.02971*, 2015.

68 X. Yu, G. Li, C. Chai, and N. Tang, "Reinforcement learning with tree-LSTM for join order selection," in *2020 IEEE 36th International Conference on Data Engineering (ICDE)*, pp. 1297–1308, IEEE, 2020.

69 I. Trummer, J. Wang, Z. Wei, D. Maram, S. Moseley, S. Jo, J. Antonakakis, and A. Rayabhari, "SkinnerDB: Regret-bounded query evaluation via reinforcement learning," *ACM Transactions on Database Systems (TODS)*, vol. 46, no. 3, pp. 1–45, 2021.

70 L. Kocsis and C. Szepesvári, "Bandit based Monte-Carlo planning," in *Machine Learning: European Conference on Machine Learning, Lecture Notes in Computer Science* (J. Fürnkranz, T. Scheffer, and M. Spiliopoulou, eds.), vol. 4212, pp. 282–293, Springer, 2006.

71 K. Tzoumas, T. Sellis, and C. S. Jensen, "A reinforcement learning approach for adaptive query processing," *History*, pp. 1–25, 2008.

72 R. Marcus, P. Negi, H. Mao, N. Tatbul, M. Alizadeh, and T. Kraska, "Bao: Making learned query optimization practical," *ACM SIGMOD Record*, vol. 51, no. 1, pp. 6–13, 2022.

73 W. R. Thompson, "On the likelihood that one unknown probability exceeds another in view of the evidence of two samples," *Biometrika*, vol. 25, no. 3–4, pp. 285–294, 1933.

Part III

Challenges, Approaches, Open Issues, and Emerging Research Topics

9

DRL Challenges in Wireless Networks

9.1 Adversarial Attacks on DRL

DRL is emerging as a potential solution with more practical applications, such as industry, agriculture, smart home, and smart cities. However, like many other ML applications, many security concerns in using DRL applications are increasingly present. Unlike conventional learning solutions, DRL is a combination of two learning algorithms, i.e. reinforcement learning and deep learning. Therefore, there are more security holes for attackers to exploit. In this section, we will discuss potential attacks on DRL-based systems, which can be divided into three main categories corresponding to the three core components of an MDP framework, i.e. action space, state space, and reward space [1]. The taxonomy of attacks on DRL models is presented in Figure 9.1.

9.1.1 Attacks Perturbing the State space

In a Markov Decision Process (MDP), the first step that an agent needs to do is to determine the current state of the system (thereby making an appropriate decision). In this way, if an attacker can somehow change the actual state of the system, it can make the agent execute decisions in favor of the attacker. This is possible in wireless communication systems where the state of the system is determined largely through sensors. For example, to determine the current state of the channel (e.g. high or low interference), the agent may need to deploy a sensor to sense the current status of the channel. In this way, if the attacker can make the agent get the wrong current state (for example, high interference instead of low interference), it can cause serious issues for the whole system. Therefore, in this section, we will explore attacks that target the state of the system using DRL and discuss some potential solutions to prevent them.

Deep Reinforcement Learning for Wireless Communications and Networking:
Theory, Applications, and Implementation, First Edition.
Dinh Thai Hoang, Nguyen Van Huynh, Diep N. Nguyen, Ekram Hossain, and Dusit Niyato.
© 2023 The Institute of Electrical and Electronics Engineers, Inc. Published 2023 by John Wiley & Sons, Inc.

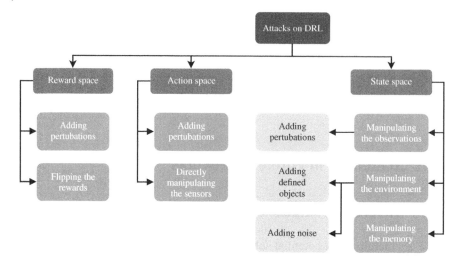

Figure 9.1 A taxonomy of attacks on DRL models. Adapted from Ilahi et al. [1].

9.1.1.1 Manipulation of Observations

In [2], the authors considered a scenario in which there is an attacker located between the agent and its interactive environment. The attacker is assumed to be able to observe the action of the agent and its states. In addition, it has the ability to change the surrounding physical environment to affect the actual state of the system. For example, in a cognitive radio network, the state of the system can be determined by the state of a target channel (i.e. idle or busy) [3]. In this way, if an attacker is placed in this environment, it can deliberately perform a signal transmission on the channel to mislead the agent that the channel is busy (even though the actual state of the channel is idle). In this way, the attacker can make the agent make decisions as expected by the attacker. However, performing such kinds of attacks is not easy. The attacker must be able to observe all states of the agent. In addition, it must be able to control the agent's states properly in order to make the agent performs as it wants. To do this, the authors in [2] proposed a smart attack based on the Fast Gradient Sign Method and Jacobian-based Salience Map Attack in order to smartly change the policy of the target via exploiting the transferability of adversarial samples. Figure 9.2 illustrates the process of such an attack with six main steps, the first four steps are at the attacker, and the last two steps are at the agent. Simulation results show that under such kind of attack, the reward function significantly reduces (approximately zero after learning epochs). Similar attacks were studied in [4], which also clearly show a significant reduction in the target's system performance.

In [5], the authors introduced two types of attacks on DRL-based systems. The first attack, namely, a strategically timed attack, aims to minimize the reward of

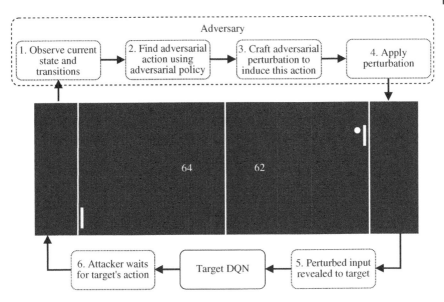

Figure 9.2 An illustration of attacks on observations of a DRL agent. Adapted from [2].

the target system by selecting an appropriate time to attack the system (instead of attacking the system during the entire learning period). It is stemmed from the fact that the efficiency of attack might be very different at different states of the system. For example, when the agent has no data packets to transmit over the channel, and at this time, if the attacker attacks the system to interrupt the communication, it will cost nothing to the system. In contrast, in case the agent has data and wants to transmit it, if the attacker attacks this time, it can significantly reduce the agent's reward. The simulation results show that such kind of attack can achieve a similar performance as that of the uniform attack, even though the number of attacks in each learning episode reduces by nearly four times. Furthermore, attacks that are launched occasionally can go undetected by the target system. The second attack introduced in this work is the enchanting attack. While the strategically timed attack aims to reduce the reward function by selecting the right state to attack, the enchanting attack aims to trick the agent into an error state that could cause the agent to make bad decisions. For example, instead of sensing that the channel is busy, the agent will be misled as the channel to be idle and transmit data onto the channel, which not only suffers from packet errors but also makes primary wireless communication systems unable to communicate and could be misunderstood as a jamming attack. Simulation results on Atari games show that the first attack can reduce the reward as that of the uniform attack, although the frequency of attack is reduced four times. Furthermore, the second attack can successfully lure the agent to the target states with a probability of more than 70%.

For wireless networks, there are some research works recently investigating security issues on states/observations of DRL frameworks. For example, in [6], the authors considered a scenario in which multiple DRL agents are using the same channel, and thus they coordinate together to find the optimal channel access and power control. Specifically, the agents will observe the status of channels and then find optimal policies to maximize the overall throughput for the whole system. However, a jammer is considered in this scenario, and this jammer is assumed to be able to make changes to the actual status of channels by jamming the channels (i.e. making the agents have wrong information/observations about the actual status of channels). In this case, the authors considered that the jammer (also controlled by DRL) could significantly degrade the system performance. A similar study on the adversarial jamming attack on multichannel access networks was also presented in [7] and [8], but with a single DRL agent. A more general case with multiple agents and multiple adversarial jamming attackers can be found in [9].

It can be observed that although there are many different types of attacks targeting states/observations of DRL systems as discussed in a recent survey [1], not many papers study these problems in wireless networks. The main reason is that there are not many attacks targeted on such networks. However, in the future, when more and more wireless systems adopting DRL frameworks, many vulnerabilities can be observed. For example, most of the current research works only focus on jamming attacks as the jammers can use their signals to make the changes in states/observations of the DRL agent (i.e. make them receive wrong information about the status of the channels). However, there are still many open problems that will need to be solved for DRL-based wireless systems. For example, in vehicular networks, the attackers can send fake information (e.g. report a wrong location) to the DRL agent. In this case, it may make the DRL agent have the wrong awareness about the surrounding environment and make a wrong decision that could cause serious consequences (e.g. crashes). Another example could be in UAV-assisted IoT data collection networks [10], where the UAV needs to check data transfer requests from IoT devices before they come to collect. In this case, if the attackers keep sending incorrect information, the DRL agent (i.e. the UAV) might make wrong decisions, and thus they can degrade the whole system's performance.

9.1.1.2 Manipulation of Training Data

Besides the attack that alters observations of DRL agents, attackers can falsify information about the states of DRL agents by transmitting false information that causes DRL agents to give incorrect data to train the deep neural networks. For example, in [11], the authors considered an attack scenario in which a Trojan is attached to a DRL agent, and it is assumed to be able to provide wrong information to the DRL agent. For example, if the DRL agent is at state s and makes an action a, it will receive a reward r. However, the attacker can change the value of reward r to r' which is very different from r. In this case, the data to train the deep neural network will be manipulated, and thus, it will make the deep neural network be trained wrongly. In this case, the attacker can control the training data, such that the optimal policy will be derived from its intention.

In terms of wireless networks, data manipulation can happen by changing sensor data. For example, a attacker can control the sensors of a DRL agent (e.g. a BS or an autonomous vehicle). Consequently, it can introduce erroneous data so that the DRL agent's training process achieves erroneous results. Or the attacker can even manipulate the training data so that the DRL agent achieves the results as the attacker's expectation. In [12], the authors considered a scenario in which a set of malicious vehicles collaborate to attack a DRL-based Adaptive Traffic Signal Control System (ATCS) via sending falsified information to make the training process at the central server go wrong (as illustrated in Figure 9.3). In particular, the authors consider an intelligent transport system using DRL to control traffic signals. The DRL agent first obtains the information from all the vehicles in its network (e.g. their locations, travel time, and traffic around the upcoming intersection) and then uses such data to train the deep neural network to find

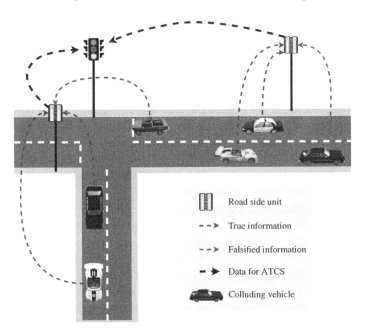

Figure 9.3 An illustration of attacks on the DRL-based adaptive traffic signal control system. Adapted from [12].

the optimal policy for the system (i.e. optimal traffic signals). However, in this network, there are a set of collected vehicles that want to collaborate to attack the system. It is assumed that malicious vehicles can collaborate and send falsified information to the DRL agent. In this way, due to training wrong information (from incorrect data sent from vehicles), the DRL agent cannot get the correct optimal policy, and thus the policy it obtains might lead to low performance. For example, instead of controlling traffics to reduce congestion on the road, it may make more traffic jams and even crashes and accidents on the road. Similar issues (i.e. manipulating training data) are also discussed in some other works such as [13–15]. They also proposed various solutions in dealing with such kinds of attacks, but the efficiency is still questionable as they still rely on the assumption about some knowledge of attackers, e.g. attackers are also using DRL algorithms with specific states, actions, and rewards parameters. However, in practice, it is nearly impossible to observe information about the attackers (e.g. their states, actions, and rewards). Alternatively, attackers in practice may want to attack randomly without following any rules and/or reasonable policies (e.g. policies and rules imposed by MDP and/or game theory). Therefore, defending against such kinds of attacks in practice is very challenging.

9.1.2 Attacks Perturbing the Reward Function

In a reinforcement learning process, an agent needs to observe the reward (e.g. the number of successfully transmitted bits) to adjust its action to achieve the final optimal policy. However, in this case, attackers can exploit the vulnerability of the system to perform attacks to degrade performance or even paralyze the network by changing information about the reward after the agent makes an action. For example, if the reward is calculated by the number of bits successfully transmitted through the response from the receiver, the attackers can falsify the information when the receiver responds to the transmitter, e.g. via a man-in-the-middle attack. In this case, the transmitter will not receive the correct feedback, and thus it will not be able to find the optimal policy for the system.

In [16], the authors investigated the performance of a software-defined networking (SDN) system using RL in case attackers are trying to attack the network's learning process by changing the actual reward the network receives. Thus, the network provider may not be able to find the optimal policy for the network. Specifically, the authors considered a scenario in which attackers are trying to propagate through the network to compromise the critical server, as illustrated in Figure 9.4. At the same time, the network agent (i.e. the RL agent) is trying to protect the critical server from compromise and preserve as many nodes as possible. To perform the attack, the authors in this work considered an attack on the reward function of the DRL agent. Specifically, after the DRL agent samples

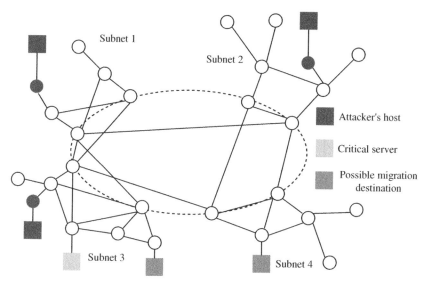

Subnet 1

Subnet 2

Attacker's host

Critical server

Possible migration
destination

Subnet 3

Subnet 4

Figure 9.4 An illustration of attacks on the reward function of the DRL agent in a software-defined network. Adapted from [16].

a batch of experiences for training, the attacker will flip the sign of a certain number of rewards (e.g. 10% of all experiences), and thus increase the reward function of the DRL agent. Note that in this scenario, the DRL agent wants to minimize its reward (loss) function. Thus, when the reward function is increased, the DRL will learn improperly. Several related works also discuss attacks on the reward function of the DRL agent, such as [17] and [18].

In the context of wireless networks, [19] is among the first works in the literature that studies the impact of reward attacks on the DRL agent. In particular, the authors considered a system model as illustrated in Figure 9.5. In this model, the base station (i.e. gNodeB) needs to allocate bandwidth resources to the mobile users (UE_1 to UE_N) based on their requirements. The BS is deployed as an RL agent that needs to make optimal decisions to maximize the long-term average reward. In particular, the RL agent's states are defined as the availability of resource blocks (RBs) and active requests from the MUs. Then, based on the observed states, the RL agent needs to assign RBs to the MUs to maximize the QoE of MUs. In this case, an adversary jammer is considered to attack the reward function of the RL agent. Specifically, the jammer also uses RL to find its best attack policy by observing RBs and choosing a number of RBs to attack. In this case, the reward of the jammer will be the number of requests that are interrupted. It is demonstrated that by attacking the reward function of the gNodeB's RL agent, the system performance could not recover even after the adversary stopped jamming.

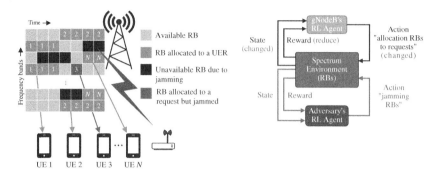

Figure 9.5 An illustration of attacks on the reward function of the DRL agent in a mobile network. Adapted from [19].

To deal with such a kind of attack, the authors in [19] proposed several countermeasures such as stopping Q-table updates when an attack is detected, introducing randomness into network slicing decisions, or manipulating the feedback mechanism in network slicing to mislead the learning process of the adversary. Simulation results show the efficiency of the proposed solutions in defending such a kind of attack. However, all these solutions are based on the assumption that we know the strategy of the attacker as well as when it attacks. In practice, these assumptions may not be realistic; therefore, more realistic solutions may need to be further developed.

9.1.3 Attacks Perturbing the Action Space

Another important factor in a DRL process that is also vulnerable to attacks is the action space of the DRL agent. In particular, when a DRL agent makes an action based on its current state and if the attacker can somehow change the taken action of the agent (the agent may not know its action is changed), then the agent may get an incorrect reward to learn. For example, if the agent is currently at state s, and it chooses an action a, then it will receive a reward r and move to state s', i.e. a tuple of $< s, a, r, s' >$ to train its DNN. However, if the attack can change the DRL's action to a', then the actual reward and next state of the agent will be different, e.g. r^* and s^*, respectively. In this case, the agent may get an incorrect experience, i.e. $< s, a, r^*, s^* >$, to learn, which can lead to a nonoptimal policy for the agent and degrade the overall system performance.

In [20], the authors introduced two types of attacks targeting the action space of a DRL agent. In particular, it is assumed that the attacker can (i) access the agent's action stream, (ii) access the agent's training environment, and (iii) know the DNN of the agent. Based on the information, the authors first design a myopic action-space (MAS) attack model which aims to make a change in the current action of the agent to minimize the immediate reward function of the agent. Although such an attack can degrade the system performance (as the agent cannot achieve the optimal policy), its efficiency is not high, as the attacker can just make some impact on the agent's immediate reward function. Therefore, the authors then introduced a more effective attack that targets minimizing the reward function of the agent in the long run. In particular, as the attack is also considered using the RL algorithm, its reward function will be designed to maximize its long-term reward (i.e. minimize the agent's long-term reward). As a result, this attack allows the attacker to achieve much higher efficiency in degrading the system performance of the agent.

Although both attacks introduced in [20] can show their impacts on attacking the DRL-based system performance, they are still constrained by some assumptions that the attacker has some advanced knowledge of the agent (e.g. can access the activity stream and know the agent's DNN) that might not be practice in real-world scenarios. Therefore, such kinds of attacks have not been widely considered in scenarios in wireless networks. However, some attacks in wireless networks, e.g. man-in-the-middle attacks, still can occur and be leveraged to launch attacks on DRL agents' actions. For example, when an IoT sends a resource request to a base station (i.e. the agent's action) via a middle node (e.g. a Wi-Fi access point), the middle node can change the request of the agent (e.g. instead of requesting 10 resource units, the middle node can change and only request two resource units from the base station). In this case, the action of the agent was intentionally changed (but the agent might not be aware of it and it might still think that it requests 10 resource units). Consequently, the agent will fail to obtain the optimal policy for the system. Many more attacks on the action space of the DRL agent are still open for research in the future of wireless networks.

9.2 Multiagent DRL in Dynamic Environments

9.2.1 Motivations

Due to the explosion of users and mobile applications, future communication networks are predicted to include many emerging networks such as vehicular, UAV, satellite communication, IoT, 5G networks, and beyond. Different from traditional mobile communication networks, the next-generation communication networks allow users to access the network almost anytime, anywhere with excellent service quality. However, they also pose many challenges that need to be addressed before these networks can be deployed in practice. In particular, these networks are often deployed on a large scale and use heterogeneous wireless communication devices, which are difficult to control. In addition, these networks are often controlled by different network entities and operated in dynamic environments with many uncertainties from the surrounding environment. In this case, single-agent DRLs (SDRLs) are often ineffective. The main reason is that the decision of one agent will be greatly influenced by the decisions of other agents working in the same environment. For example, when deciding to access a channel, after a learning process from the environment, agent A may decide to access the channel with probability p_1. However, at that time, there is another agent, i.e. agent B, who also wants to learn to access the channel. Therefore, agent B's channel access decisions will change the learning environment, leading to incorrect learning decisions from agent A. As a result, neither agents A nor B will be able to find the decisions optimal, and even their performance after the learning process is reduced due to conflicts of learning purposes. Therefore, Multi-agent Deep Reinforcement Learning (MDRL) algorithms [21, 22] have been introduced recently as effective solutions to solve problems in wireless communication networks. In the following, we will study two typical MDRL models used in wireless communication networks.

9.2.2 Multiagent Reinforcement Learning Models

As in MDRL, we consider scenarios with multiple DRL agents, and there are two typical scenarios used in wireless communication networks. The first scenario is when DRL agents belong to different network entities, and thus, they usually have different objectives. For example, DRL agents may only want to minimize their own energy consumption (without considering others' energy consumption). In this case, Markov game models (or Stochastic game models) can be used to formulate and find the equilibrium for the agents. In the second scenario, when all the agents share the same objective function (e.g. all the agents belong to the same network entity, and they want to do the same work together), we can use the Decentralized Partially Observable Markov Decision Processes (DPOMDP) models to formulate and find the jointly optimal policy for the whole network.

In general, MDRL offers several advantages over SDRL:

- MDRL can solve complex optimization problems in wide-area wireless communication networks that are often NP-hard or nonconvex (e.g. interference management for hierarchical multicell networks and power control for heterogeneous wireless devices). For example, instead of solving a large-scale optimization problem (e.g. consisting of many cells), we can break it down into small problems so that the DRL agents can learn simultaneously. In this way, we can not only reduce the complexity of the global optimization problem but also speed up the process of finding the optimal value for the optimization problem.
- Unlike SDRL algorithms, MDRL algorithms allow agents to not only learn the influences of the surrounding environment but also learn the actions of other

agents by observing and evaluating their impacts. Therefore, the obtained optimization policy can optimize performance for itself and for other agents.

- One of the outstanding advantages of MDRL is that it allows to control and manage the communication network in a distributed, efficient, and flexible manner. Specifically, with MDRL, decisions will be made by DRL agents simultaneously in real time based on their own observations and policies. This thus avoids the delays and bottlenecks that often occur in centralized network control.

- During the learning process, MDRL models allow agents to collaborate and share data, thereby speeding up the learning process as well as reducing communication and computation costs. For example, when an agent (e.g. a UAV) leaves the group and another one comes to replace it. Instead of learning from the beginning, other agents (e.g. UAVs) in the networks can share their information (e.g., trained neural networks) with the new UAV, so that it can learn the optimal policy quickly.

- MDRL models also offer flexible strategies for service providers. For example, as discussed above, in cases where the DRL agents belong to the same network entity and their aims are to maximize the joint reward function, DPOMDP can be used as an effective solution. In contrast, if they belong to different entities and have dissimilar objectives, a Markov game model can be useful to analyze and find their equilibrium.

In the following, we will discuss these models in detail.

9.2.2.1 Markov/Stochastic Games

Markov games (or stochastic games) are an extension of MDP, in which DRL agents (formulated by MDP) work and interact in the same environment [23, 24]. Let us denote $\mathcal{N} = \{1, \ldots, N\}$ as a set of agents who are interacting in the same environment. All the agents have the same state space S, but they may have different action spaces, which are denoted by $\mathcal{A}_n$, where $n \in \mathcal{N}$.

At each time slot t, all the agents observe the same state s ($s \in S$), and then each agent n will select an action a_n based on its own policy π_n. When all the actions are executed, the system will move to a new state with the following transition probability function $\mathcal{P} : S \times \mathcal{A} \times S \to [0, 1]$. Here, $\mathcal{A}$ is the joint action space which can be defined as a combination of all the action spaces from agents, i.e. $\mathcal{A} = \mathcal{A}_1 \times \mathcal{A}_2 \times \cdots \times \mathcal{A}_N$. At the same time, each agent n will receive an immediate reward $\mathcal{R}_n = S \times \mathcal{A} \times S \to \mathbb{R}$.

In this way, a Markov game can be defined as a tuple of $(N, S, (\mathcal{A}_n)_{n \in \mathcal{N}}, \mathcal{P}, (\mathcal{R}_n)_{n \in \mathcal{N}}, \gamma)$, where γ is a discount factor that is used to evaluate the future reward. Based on this formulation, each agent will try to find its optimal policy π_n^* in order to maximize its long-term reward. If we define π_{-n} as the joint policy of all the agents except agent n, then the value function of an agent n can be defined as follows:

$$V^n_{\pi_n,\pi_{-n}}(s) := \mathbb{E}\left[\sum_{t=0}^{\infty}\gamma^t R^n(s_t, a_t, s_{t+1})\big|_{a_t^n \sim \pi^n(.|s_t), s_0 = s}\right]. \tag{9.1}$$

Based on this value function, we can observe that the optimal policy of agent n will be impacted largely by other agents' policies. This is the main reason that causes difficulties in finding the optimal policy for the agents as their policies are highly dynamic and uncertain during the learning process (e.g. due to impacts by the environments and other agents' actions).

In Markov games, there are two models that are widely adopted in wireless networks.

- *Fully competitive*: In this case, each agent has its own reward function, and the main objective of each agent is to maximize its own reward function without considering other agents' reward functions. This is a typical scenario used in the context of Markov games in wireless networks. In special cases when the sum of all the agents' reward functions is equal to zero, i.e. $\sum_{n=1}^{N} R_n = 0$, this is known as a zero-sum game [25], a very strong competitive game as the positive reward function has a direct negative impact to others' reward functions.
- *Fully cooperative*: This is a special case when all the agents share the same reward function. This case is also known as the multiagent MDP [26]. This scenario only happens when all the agents belong to and are controlled by a network entity. This case can be solved straightforwardly by conventional single DRL algorithms:

9.2.2.2 Decentralized Partially Observable Markov Decision Process (DPOMDP)

A DPOMDP is also widely used to formulate dynamic decision-making problems for multiple DRL agents in the same environment. Similar to the special case of fully cooperative Markov games, the agents in the DPOMDP also share the same reward function, and they aim to maximize this joint reward function after the learning process. However, there is one major difference between DPOMDP and Markov game models. Specifically, in a Markov game model, all the agents share the same state, and it is assumed that all the agents can observe the global state of the system (i.e. s) at every time slot. However, in a DPOMDP model, each agent has its own state space, i.e. $S_n, \forall n \in \mathcal{N}$, and at each time slot, each agent only can make a decision based on its own observation o_n. Here, $o_n \in \mathcal{O}_n$ in which $\mathcal{O}_n$ is the observation space of the agent n. This is the most difficult part of the DPOMDP since each agent only has a part of the whole system state, and it needs to learn and explore all other activities and states to find its own optimal policy. The comparisons among different models can be seen in Figure 9.6.

Figure 9.6 Comparisons among different Markov models. (a) MDP, (b) POMDP, (c) Markov games, and (d) Dec-POMDP.

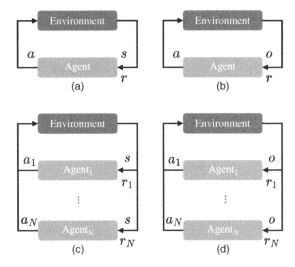

In this way, we can define a DPOMDP as a tuple of $(N, S, (\mathcal{A}_n)_{n \in \mathcal{N}}, P, R, \gamma,$ $(\mathcal{O}_n)_{n \in \mathcal{N}}, Z)$ where most factors are the same as defined in the Markov game above. There are only two new factors here including $\mathcal{O}_n$ and $\mathcal{Z} : S \times \mathcal{A} \times \mathcal{O} \rightarrow [0, 1]$ which are the agent n's observation space and the joint observation function, respectively. In this model, at each time slot t, each agent-n observes its current observation, executes an action from its action space, and then observes a new observation based on the following transition probability function $P(\mathcal{O}|\mathcal{A}, s')$.

Here, it is noted that, if all the agents have different reward functions, then the DPOMDP model is also known as Partially Observable Stochastic Games [27]. This game model is the most complicated case in multiagent reinforcement learning as the agents need to learn in a competitive environment with very little information from the surrounding environment as well as from other competitors (i.e. agents).

9.2.3 Applications of Multiagent DRL in Wireless Networks

As an emerging topic in wireless networks, multiagent DRL has a large number of applications in wireless networks. As shown in a recent survey on this topic [21], there is a wide range of problems that MDRL can be used to address in wireless networks. Specifically, as shown in Figure 9.7, MDRL algorithms can be adopted to address diverse problems, such as network access, power control, offloading, caching, packet routing. In addition, Figure 9.7 shows the percentages of publications of MDRL in different issues as well as different wireless networks (e.g. IoT, cellular, UAV, and vehicle networks). All these numbers can provide an overview of information and potential applications of MDRL in future wireless networks.

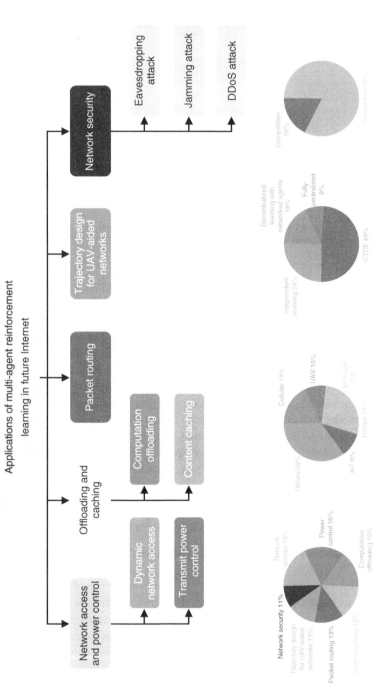

Figure 9.7 Classification of applications of multiagent DRL in wireless networks. Adapted from [21].

9.2.4 Challenges of Using Multiagent DRL in Wireless Networks

Despite possessing many advantages with many potential applications, the implementation of MDRL models in wireless networks is facing many difficulties and challenges. In this section, we will discuss two major challenges in using MDRL in real-world wireless communication networks.

9.2.4.1 Nonstationarity Issue

In MDRL, the DRL agents have to learn and make actions at the same time. This may make a lot of issues in the learning process. This is because when a DRL agent makes an action, its action has influenced not only the surrounding environment but also the decision of other DRL agents who are interacting and learning in the same environment. This is known as the nonstationarity issue in MDRL [28], which is also the key difference between SDRL and MDRL processes. Due to the nonstationarity issue, the convergence of MDRL algorithms for each learning agent cannot be guaranteed. To deal with this problem, there are several solutions proposed recently, e.g. centralized critic techniques, opponent modeling, and meta-learning. For more detailed information, interested readers can refer to this survey [29].

9.2.4.2 Partial Observability Issue

This is a typical issue yet a big challenge in MDRL. In MDRL, due to decentralized control models, each DRL agent has to make actions individually based on all the information it can observe from the surrounding environment. Unfortunately, unlike SDRL, the information it observes is very limited. For example, the agent may know the change in the environment after it makes a decision but has no idea of the states and actions of the other agents in the same environment. This is the main issue that makes the learning processes not only much slower than those of SDRL algorithms but also very unstable and even cannot converge. To address this problem, one of the most popular approaches is exploring communications among agents in the network. For example, the agents can communicate and are under control by a centralized node during the learning process (also known as fully centralized learning), or the agents can be trained in a centralized manner at a central node and then execute actions in a decentralized fashion (also known as centralized training with decentralized execution). However, in wireless networks, communications among the agents and/or between the agents and centralized nodes are not easy to maintain due to the dynamic and uncertainty of the wireless environments. More interesting solutions can be found in this survey article [30, 31].

9.3 Other Challenges

9.3.1 Inherent Problems of Using RL in Real-Word Systems

In addition to the above challenges and open issues, this section will focus on some inherent problems encountered when deploying RL algorithms in real-work systems. In practice, when these algorithms are implemented and executed, many new issues can happen [32]. If they are not aware of and handled in time, they can cause other issues that not only affect system performance but can also have serious consequences, e.g. security and human health, especially in some areas such as smart cities, healthcare, and military communication networks.

9.3.1.1 Limited Learning Samples

From most of the literature on this topic presented in some recent surveys (e.g. [21, 33]), we can observe that most of the current works assume that they have a perfect simulation environment for the DRL agent to learn and interact. Due to the simulation environment, training data (or experiences) can be obtained quickly and diversely to serve the training process, and thus the optimal policy can be obtained quickly (e.g. even after just a few thousand epochs). Unfortunately, this is not the case in practice. In real systems, the DRL agent may have very limited qualified data samples to learn. Most of the collected data are repeated and do not provide sufficient information for the agents to effectively explore the surrounding environment, especially at the beginning of the learning period. For example, when a wireless device observes the channel to learn its traffic demand, it would be very likely that the information it observed is very similar and might not be useful to learn within a short period of time. To address this problem, transfer learning techniques [34] could be an effective solution, e.g. to transfer some similar knowledge from similar agents learning in similar environments, thereby overcoming the lack of learning samples of reinforcement learning algorithms in practice. Interesting readers can refer to the recent survey on this topic for further information [34].

9.3.1.2 System Delays

In most of the current research works in the literature, they are assumed that the agent can always obtain the current state. Once it makes a decision, it can immediately obtain its reward (all events are assumed to be within a one-time slot). However, in real-work wireless communication systems, it might not be the case. For example, to obtain the channel state, the wireless device may need to sense the channel to obtain the information (e.g. the channel is idle or busy), it usually takes time to get the correct information. In addition, when the agent (e.g. wireless device) makes an action, e.g. transmitting data, it might not be able to immediately obtain the result (e.g. how many packets were successfully transmitted). It may take time to receive feedback from the receiver, and even in some cases, the feedback from the receiver is delayed or lost due to poor wireless connections. All these cases can happen at any time due to the uncertainty of the wireless environments. If they happen, the algorithm may not work properly, and even break down the whole system.

9.3.1.3 High-Dimensional State and Action Spaces

This is also a fairly common problem in research. In many papers, the system states and the actions of the agent are often simplified to make the system easier to formulate and find the optimal policy. For example, energy states are often assumed to be broken down into discrete energy levels with a finite number of states (e.g. energy levels). Likewise, actions (e.g. transmit power) are also assumed to be broken down into small levels of effort with a finite number of levels so that wireless developers can choose to transmit. However, this is very unlikely to happen in real wireless communication systems. Even if we can assume that states and actions can be discrete, their numbers can even reach several thousand or even a few million units (due to high-dimensional issues). Therefore, normalizing and formulating action and state spaces from real problems is also one of the big challenges in using DRL algorithms.

9.3.1.4 System and Environment Constraints

Realistic communication systems are always limited by constraints, such as bandwidth, power, energy, and processing speed. Some problems can be transformed into system states, e.g. a UAV may have limited energy, and in this case, the energy of the UAV can be converted to the system state with maximum energy (as the limit of energy capacity) to be the final state of the system [10]. However, many problems in practice have very complex constraints which are difficult to formulate in terms of system states. For example, to ensure the quality of service for users, communication delay (which is a function of the transmitted power and allocated bandwidth) must be less than a predetermined limit (e.g. $delay = \frac{f(power)}{f(bandwidth)} \leq \tau$). However, constraints are difficult to transform into system states to return to traditional MDP forms. Therefore, solutions using constrained MDP [35] can be considered.

9.3.1.5 Partial Observability and Nonstationarity

These are also two fairly common problems we face when implementing MDPs in real wireless communications systems. First, in order to make optimal decisions, we need to get a comprehensive view of the system. This requires a large number of sensors that are able to accurately collect and analyze data from the system and its surroundings. However, this is very difficult to do due to physical barriers as well as implementation costs. Therefore, for most real systems, we can only observe a part of the system (or partially observable) to make decisions. Second, unlike the simulated environment, the real environment has a lot of noise and uncertainties that are almost unrealistic to model as a random function (e.g. Gaussian noise function). Besides the factors caused by the interference of the transmission medium, there are many other types of noise and uncertainties that are difficult to predict and model, e.g. hardware, software, and even human impact factors (e.g. system manipulation errors or unusual interventions during operation). All these factors have caused a lot of difficulties in applying and implementing DRL algorithms in practice.

9.3.1.6 Multiobjective Reward Functions

Although this is not a common problem in wireless communication control systems using DRL, it also causes certain difficulties for the implementation of DRL algorithms. In particular, this problem often occurs when a DRL agent wants to optimize multiple goals at the same time. For example, the agent (e.g. a computing service provider) wants to not only maximize its profit from providing mobile computing offloading services but also minimize the total cost to maintain the service for the client users. These objective functions often conflict and have different units, so they cannot be combined straightforwardly into a joint global objective function. Therefore, one of the popular methods is to use weights to normalize these functions to the same unit so they can be combined via a sum function. For example, $global\ function = weight_1 \times objective\ function_1 + weight_2 \times objective\ function_2$. However, the downside of using weights is that the weights are quite difficult to control, and sometimes they cannot truly reflect the agent's goal.

There is a recent comprehensive survey that thoughtfully studies the challenges of real-world reinforcement learning [32], and interesting readers can refer to this article for more detailed information.

9.3.2 Inherent Problems of DL and Beyond

Section 9.3.1 discusses issues mainly related to the implementation of DRL algorithms in wireless communication systems from the perspective of reinforcement learning algorithms, this section discusses aspects of deep learning algorithms in more depth. We first review the general issues of the DL algorithm in practical implementation. We then explore the challenges and issues of implementing DL in DRL algorithms in real-world wireless communication systems.

9.3.2.1 Inherent Problems of DL

As DRL is a combination of DL and RL, it will inherit all problems that can occur when DL is implemented in practice. In the following, we list some typical issues that can happen in implementing DL in DRL-based problems in wireless networks. Interesting readers can refer to the following articles for more detailed information [36–38]:

- The key to achieving good performance when using DL algorithms is that we need to collect a large amount of data for training. However, this is also a big issue for DRL agents because agents need a fair amount of time to collect enough information to train deep neural networks.
- The second problem is data labeling. Usually, deep neural networks need prelabeled data to be effective in learning. However, when DRL agents collect data from the surrounding environment, it is sometimes difficult for the data to be classified and labeled correctly (e.g. due to disturbances or uncertainties from the wireless environments). For example, a data packet that has been transmitted unsuccessfully can be mistaken for successful transmission. Therefore, that data will be mislabeled, leading to the training process of deep neural networks not being as effective as it should be desired.
- Another problem that also frequently occurs in the implementation of DL algorithms in practice is that the data contain very little information and does not have much value to learn. For example, when a UAV is on a mission to fly around a certain area to collect data; however, most of the time, the UAV only flies in certain areas, e.g. around a shopping mall. However, the UAV may not often go to other areas to collect data. Therefore, the amount of data obtained is unbalanced and can be biased when training its deep neural network.
- Training large amounts of data usually requires a large number of computational resources to perform. However, wireless devices (e.g. smartphones, IoT devices, and UAVs) are often limited in hardware and power, and thus they are difficult to perform intensive computing tasks.
- Unlike RL, DL has no strong theory to prove the optimality of actions and policies. All results obtained from DL algorithms are relative and very dependent on adjusting the learning parameters of the network, such as the number of neural networks, the number of layers, the weights of the nodes. Therefore, the design and implementation of DLs in DRL-based problems in practice require skilled engineers who have rich experiences in both DL and wireless networks.

9.3.2.2 Challenges of DRL Beyond Deep Learning

Nonstationary Target Function Besides the general problems when using DL, there are some specific problems when using DL in DRL algorithms. Let us take a look at a typical code for DRL algorithm as in Algorithm 9.1. As observed in the above code on lines 11 and 12, unlike conventional DL algorithms, in each iteration of the DRL algorithm, the target function can be changed depending on the states and actions that the agent observes and performs. The reason is that at the beginning of the learning period, the agent has no idea which ones are the best Q-values it should take, and thus it needs to explore (by taking some random actions based on greedy policy to learn which ones are the best). This is the fundamental difference between DL and DRL, and it also leads to the nonstationary issue during the training of DRL algorithms.

Algorithm 9.1 Deep Q-learning algorithm with experience replay.

1: Start with $Q_0(s, a)$ for all s and a.
2: Get initial state s_0
3: **for** $k=1,2,...$ till converge **do**
4:      Sample action a, get next state s'
5:      **if** s' is terminal: **then**
6:          target $= R(s, a, s')$
7:          Sample new initial state s'
8:      **else**
9:          target $= R(s, a, s') + \gamma \max_{a'} Q_k(s', a')$
10:     **end if**
11:     $\theta_{k+1} \leftarrow \theta_k - \alpha \nabla_\theta \mathbb{E}_{s' \sim P(s'|s,a)}[(Q_\theta(s, a) \text{ - target}(s'))^2]|_{\theta=\theta_k}$
12:     $s \leftarrow s'$
13: **end for**

To overcome this problem, one of the most popular solutions is to use two deep neural networks for training at the same time, instead of one deep neural network like conventional DRL algorithms. Specifically, as shown in Figure 9.8, we can use two deep neural networks called the target network and the prediction network. The parameters of the target network will be fixed for a specific period of time, e.g. C iterations, while the parameters of the prediction network will be updated regularly. After every C iterations, all parameters of the prediction network will be copied to the target network for learning. As a result, this solution allows the training process more stable as the target network will be fixed during every C iterations. However, controlling the number of iterations C is not easy and usually requires hands-on skills and a lot of trials (as it much depends on the system configurations) to find optimal values.

High-Correlation Experiences   Another distinctive feature we can observe here is the correlation of samples in DRL algorithms. In particular, unlike DL algorithms, DRL algorithms obtain samples (i.e. experiences) in a real-time manner (i.e. an agent makes an action a at state s, then observes a new state s' and receives an immediate reward r, all these information $< s, a, s', r >$ makes a new sample for

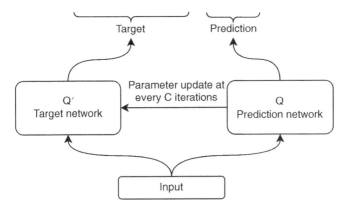

Figure 9.8 An illustration of using two separated deep neural networks, i.e. target network and prediction network, for the learning process.

Figure 9.9 An illustration of DRL agent using reply memory to mitigate high-correlation of collected experiences.

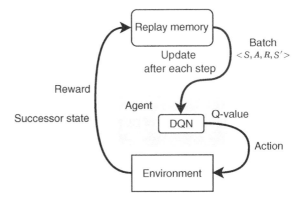

the deep neural network to train). As samples are obtained in this way, they might be very similar and highly correlated. Consequently, the learning performance will be very low when using a deep neural network as the samples do not have sufficient and diverse information to learn.

To address this issue, one of the most popular methods is replaying experiences. Specifically, a memory pool is used to store all the experiences when the agent interacts with the surrounding environment (as illustrated in Figure 9.9). A batch of experiences is randomly selected from the memory pool and used for training the deep neural network. The batch of experiences will be renewed regularly to make sure that the agent has new and diverse data to learn from. This enhances the learning performance of the agent. However, how to design the batch size and how frequently it should be updated are important issues for practical implementation.

Table 9.1 Deep learning frameworks and their supported hardware platforms.

	CPU -x86	CPU -Arm	GPU -CUDA	GPU -OpenCL	APU	NVDLA	NPU	Specific ASICs	FPGA
TensorFlow	✓		✓	✓					
TensorFlow lite	✓	✓	✓	✓				✓	
PyTorch	✓	✓	✓	✓				✓	
PyTorch mobile	✓	✓	✓	✓				✓	
Tencent Neural Network (TNN)	✓	✓		✓			✓		
MiniDNN		✓		✓	✓				
Caffe	✓	✓	✓	✓				✓	✓
Apple CoreML		✓						✓	
TebsorRT	✓	✓	✓			✓			
Apache TVision Machine Learning (TVM)	✓	✓	✓	✓				✓	✓
Xiaomi MACE	✓	✓		✓	✓				

9.3.3 Implementation of DL Models in Wireless Devices

In most of the current papers using DRL algorithms, they only consider simulations, so implementing deep neural networks is quite simple. However, in practice, running deep neural networks on wireless devices faces many difficulties which require the optimization and standardization of both software and hardware of wireless devices.

In [39], the authors conducted a survey and synthesized many deep learning frameworks and their corresponding compatible hardware platforms that can be used on mobile devices, as summarized in Table 9.1. As observed in this table, both CPU and GPU hardware platforms very well support running most of the deep learning frameworks. Alternatively, specific ASICs (e.g. Apple Bionic, Google TPU) platforms can support some light versions of deep learning frameworks such as TensorFlow Lite, PyTorch. However, nearly half of hardware platforms are still not ready to support running deep learning algorithms, e.g. AI processing unit (APU), NVIDIA deep learning accelerator (NVDLA), neural processing unit (NPU), and field-programmable gate array (FPGA).

Here, it is to be noted that although there are quite many hardware platforms that can support running deep learning algorithms, deploying them on wireless devices is not straightforward. For example, when deploying deep learning frameworks on wireless devices running iOS and Android platforms, they need to be compatible with both their hardware platforms and operating systems.

9.4 Chapter Summary

We have discussed important challenges and open issues in deploying DRL applications in wireless networks in practice. There are diverse challenges and open issues when implementing DRL algorithms in practice, e.g. adversarial attacks on DRL algorithms, multiagent DRL in dynamic environments, limited learning samples, real-time delay, and hardware constraints of wireless devices. These issues and challenges need to be carefully addressed before widespread deployment of DRL algorithms in wireless networks.

References

1 I. Ilahi, M. Usama, J. Qadir, M. U. Janjua, A. Al-Fuqaha, D. T. Hoang, and D. Niyato, "Challenges and countermeasures for adversarial attacks on deep reinforcement learning," *IEEE Transactions on Artificial Intelligence*, vol. 3, pp. 90–109, 2022.

2 V. Behzadan and A. Munir, "Vulnerability of deep reinforcement learning to policy induction attacks," in *International Conference on Machine Learning and Data Mining in Pattern Recognition*, pp. 262–275, 2017.

3 D. T. Hoang, D. Niyato, P. Wang, and D. I. Kim, "Opportunistic channel access and RF energy harvesting in cognitive radio networks," *IEEE Journal on Selected Areas in Communications*, vol. 32, no. 11, pp. 2039–2052, 2014.

4 V. Behzadan and A. Munir, "Adversarial attacks on neural network policies," in *ICLR Workshop*, pp. 1–7, 2017.

5 Y. C. Lin, Z. W. Hong, Y. H. Liao, M. L. Shih, M. Y. Liu, and M. Sun, "Tactics of adversarial attack on deep reinforcement learning agents," in *Proceedings of the 26th International Joint Conference on Artificial Intelligence*, pp. 3756–3762, 2017.

6 F. Wang, M. C. Gursoy, and S. Velipasalar, "Adversarial reinforcement learning in dynamic channel access and power control," in *IEEE Wireless Communications and Networking Conference*, pp. 1–6, 2021.

7 C. Zhong, F. Wang, M. C. Gursoy, and S. Velipasalar, "Adversarial jamming attacks on deep reinforcement learning based dynamic multichannel access," in *IEEE Wireless Communications and Networking Conference*, pp. 1–6, 2020.

8 F. Wang, C. Zhong, M. C. Gursoy, and S. Velipasalar, "Resilient dynamic channel access via robust deep reinforcement learning," *IEEE Access*, vol. 9, pp. 163188–163203, 2021.

9 J. Dong, S. Wu, M. Sultani, and V. Tarokh, "Multi-agent adversarial attacks for multi-channel communications," *arXiv preprint arXiv:2201.09149*, 2022.

10 N. H. Chu, D. T. Hoang, D. N. Nguyen, N. V. Huynh, and E. Dutkiewicz, "Joint speed control and energy replenishment optimization for UAV-assisted IoT data collection with deep reinforcement transfer learning," *IEEE Internet of Things Journal*, Early Access, 2022.

11 K. Panagiota, W. Kacper, S. Jha, and L. Wenchao, "TrojDRL: Trojan attacks on deep reinforcement learning agents," in *Proceedings of the 57th ACM/IEEE Design Automation Conference (DAC)*, 2020.

12 A. Qu, Y. Tang, and W. Ma, "Attacking deep reinforcement learning-based traffic signal control systems with colluding vehicles," *arXiv preprint arXiv:2111.02845*, 2021.

13 A. Talpur and M. Gurusamy, "Adversarial attacks against deep reinforcement learning framework in internet-of-vehicles," in *2021 IEEE Globecom Workshops (GC Wkshps)*, pp. 1–6, IEEE, 2021.

14 A. Ferdowsi, U. Challita, W. Saad, and N. B. Mandayam, "Robust deep reinforcement learning for security and safety in autonomous vehicle systems," in *2018 21st International Conference on Intelligent Transportation Systems (ITSC)*, pp. 307–312, IEEE, 2018.

15 M. Li, Y. Sun, H. Lu, S. Maharjan, and Z. Tian, "Deep reinforcement learning for partially observable data poisoning attack in crowdsensing systems," *IEEE Internet of Things Journal*, vol. 7, no. 7, pp. 6266–6278, 2019.

16 Y. Han, B. I. Rubinstein, T. Abraham, T. Alpcan, O. D. Vel, S. Erfani, D. Hubczenko, C. Leckie, and P. Montague, "Reinforcement learning for autonomous defence in software-defined networking," in *Decision and Game Theory for Security. International Conference on Decision and Game Theory for Security, Lecture Notes in Computer Science* (L. Bushnell, R. Poovendran, and T. Başar, eds.), vol. 1199, pp. 145–165, Springer, 2018.

17 Y. Huang and Q. Zhu, "Deceptive reinforcement learning under adversarial manipulations on cost signals," in *Decision and Game Theory for Security. International Conference on Decision and Game Theory for Security, Lecture Notes in Computer Science* (T. Alpcan, Y. Vorobeychik, J. Baras, and G. Dán, eds.), vol. 11836, pp. 217–237, Springer, 2019.

18 A. Rakhsha, G. Radanovic, R. Devidze, X. Zhu, and A. Singla, "Policy teaching via environment poisoning: Training-time adversarial attacks against reinforcement learning," in *International Conference on Machine Learning*, pp. 7974–7984, PMLR, 2020.

19 Y. Shi, Y. E. Sagduyu, T. Erpek, and M. C. Gursoy, "How to attack and defend 5G radio access network slicing with reinforcement learning," *arXiv preprint arXiv:2101.05768*, 2021.

20 X. Y. Lee, S. Ghadai, K. L. Tan, C. Hegde, and S. Sarkar, "Spatiotemporally constrained action space attacks on deep reinforcement learning agents," *Proceedings of the AAAI Conference on Artificial Intelligence*, vol. 34, pp. 4577–4584, 2020.

21 T. Li, K. Zhu, N. C. Luong, D. Niyato, Q. Wu, Y. Zhang, and B. Chen, "Applications of multi-agent reinforcement learning in future internet: A comprehensive survey," *IEEE Communications Surveys & Tutorials*, vol. 24, no. 2, pp. 1240–1279, 2022.

22 A. Feriani and E. Hossain, "Single and multi-agent deep reinforcement learning for AI-enabled wireless networks: A tutorial," *IEEE Communications Surveys & Tutorials*, vol. 23, no. 2, pp. 1226–1252, 2021.

23 L. S. Shapley, "Stochastic games," *Proceedings of the National Academy of Sciences of the United States of America*, vol. 39, no. 10, pp. 1095–1100, 1953.

24 M. L. Littman, "Markov games as a framework for multi-agent reinforcement learning," in *Machine learning proceedings 1994*, pp. 157–163, Elsevier, 1994.

25 M. G. Lagoudakis and R. Parr, "Learning in zero-sum team Markov games using factored value functions," *Advances in Neural Information Processing Systems 15*, 2002.

26 C. Boutilier, "Sequential optimality and coordination in multiagent systems," *IJCAI*, vol. 99, pp. 478–485, 1999.

27 E. A. Hansen, D. S. Bernstein, and S. Zilberstein, "Dynamic programming for partially observable stochastic games," *AAAI*, vol. 4, pp. 709–715, 2004.

28 P. Hernandez-Leal, M. Kaisers, T. Baarslag, and E. M. de Cote, "A survey of learning in multiagent environments: Dealing with non-stationarity," *arXiv preprint arXiv:1707.09183*, 2017.

29 G. Papoudakis, F. Christianos, A. Rahman, and S. V. Albrecht, "Dealing with non-stationarity in multi-agent deep reinforcement learning," *arXiv preprint arXiv:1906.04737*, 2019.

30 W. Du and S. Ding, "A survey on multi-agent deep reinforcement learning: From the perspective of challenges and applications," *Artificial Intelligence Review*, vol. 54, no. 5, pp. 3215–3238, 2021.

31 L. Canese, G. C. Cardarilli, L. Di Nunzio, R. Fazzolari, D. Giardino, M. Re, and S. Spanò, "Multi-agent reinforcement learning: A review of challenges and applications," *Applied Sciences*, vol. 11, no. 11, p. 4948, 2021.

32 G. Dulac-Arnold, N. Levine, D. J. Mankowitz, J. Li, C. Paduraru, S. Gowal, and T. Hester, "Challenges of real-world reinforcement learning: Definitions, benchmarks and analysis," *Machine Learning*, vol. 110, no. 9, pp. 2419–2468, 2021.

33 N. C. Luong, D. T. Hoang, S. Gong, D. Niyato, P. Wang, Y. C. Liang, and D. I. Kim, "Applications of deep reinforcement learning in communications and networking: A survey," *IEEE Communications Surveys & Tutorials*, vol. 21, no. 4, pp. 3133–3174, 2019.

34 C. T. Nguyen, N. Van Huynh, N. H. Chu, Y. M. Saputra, D. T. Hoang, D. N. Nguyen, Q.-V. Pham, D. Niyato, E. Dutkiewicz, and W.-J. Hwang, "Transfer learning for wireless networks: A comprehensive survey," *Proceedings of the IEEE*, vol. 110, no. 8, pp. 1073–1115, 2022.

35 E. Altman, *Constrained Markov Decision Processes: Stochastic Modeling*. Routledge, 1999.

36 I. Goodfellow, Y. Bengio, and A. Courville, *Deep Learning*. MIT Press, 2016.

37 C. Zhang, P. Patras, and H. Haddadi, "Deep learning in mobile and wireless networking: A survey," *IEEE Communications Surveys & Tutorials*, vol. 21, no. 3, pp. 2224–2287, 2019.

38 Q. Mao, F. Hu, and Q. Hao, "Deep learning for intelligent wireless networks: A comprehensive survey," *IEEE Communications Surveys & Tutorials*, vol. 20, no. 4, pp. 2595–2621, 2018.

39 H. Tabani, A. Balasubramaniam, E. Arani, and B. Zonooz, "Challenges and obstacles towards deploying deep learning models on mobile devices," *arXiv preprint arXiv:2105.02613*, 2021.

10

DRL and Emerging Topics in Wireless Networks

10.1 DRL for Emerging Problems in Future Wireless Networks

10.1.1 Joint Radar and Data Communications

Joint Radar and Data Communication (JRDC) [1, 2] has recently been emerging as a novel technology that is widely used in wireless networks with a wide range of applications, especially in autonomous vehicle systems. The basic principle of JRDCs is to allow wireless transmission devices to use two functions, i.e. data transmission and radar detection, at the same time on the same frequency. In this way, we can maximize the use of bandwidth resources that have gradually been depleted due to the explosion of mobile devices and wireless communication services. In addition, allowing two communication functions (i.e. radar and data communications) to work together and support each other will open a number of new applications with a lot of benefits to human beings, including both military applications (e.g. ship-borne, airborne, and ground-based JRDC systems) and civilian applications (e.g. autonomous vehicular systems, Wi-Fi communications integrated with indoor localization and activity recognition, and joint UAV communications and radar functions) [1, 2].

Among various applications of JRDC, autonomous vehicles (AVs) are known as the most successful and also the most popular application in using JRDC at the present time. This is due to the following two main reasons. First, in current Intelligent Transportation Systems (ITSs), data communication between AVs as well as between AVs and infrastructure is essential to enable AVs and ITSs to timely provide and continuously update information for users, thereby providing safe and enjoyable experiences during their journeys. This is also an important reason for the strong development of vehicular networks worldwide. The second reason is due to the safety issue of using AVs. In fact, AVs are increasingly being developed to allow them to operate automatically and to be

Deep Reinforcement Learning for Wireless Communications and Networking:
Theory, Applications, and Implementation, First Edition.
Dinh Thai Hoang, Nguyen Van Huynh, Diep N. Nguyen, Ekram Hossain, and Dusit Niyato.
© 2023 The Institute of Electrical and Electronics Engineers, Inc. Published 2023 by John Wiley & Sons, Inc.

less dependent on user controls. Although this can reduce the dependence on drivers, it imposes many safety-related requirements on the use of AVs. Therefore, radar communication systems have been developed and widely deployed on most of the AVs to provide better information about their surroundings (especially information/objects that cannot be observed with the human eyes or cannot be detected through conventional sensors, e.g. cameras). In fact, several companies such as NXP (www.nxp.com), Rohde&Schwarz (www.rohde-schwarz.com), and Infineon (www.infineon.com) have recently introduced new generation radar systems that allow detecting and recognizing objects at distances up to 250 m with very high accuracy. Therefore, the combination of data communication and radar communication (or JRDC) is considered a great development step for the development of AVs in the near future.

In AVs using JRDC systems, as they use two functions at the same time, i.e. transmitting data or detecting objects in the surrounding environment, on the same hardware device, they have to share a lot of common resources such as spectrum, power, and antenna. Therefore, a fundamental question here is how to use these two functions most effectively, given that they share the same hardware resources on the same device. In addition, another big challenge when using the functions of JRDC systems on AVs is how to use them effectively in a real-world environment with lots of dynamics and uncertainty. Specifically, when an AV is on the road, they can enter dense areas with many obstacles blocking the driver's view, or it could be in an environment with extreme conditions, for example heavy rain or snow. In this case, the radar function must be used more often to increase the driver's safety. In the opposite case, e.g. when the AV is in a favorable condition (e.g. good weather conditions and a wide-open area) and it has a lot of data to transmit. The data transfer function can be prioritized and allocated more resources in this case. Therefore, to provide a flexible solution in a dynamic and uncertain environment, DRL can be used as a suitable solution to solve this problem.

In [3], the authors proposed a framework that uses MDP and DRL to enable an AV to make optimal dynamic decisions based on its current state, as depicted in Figure 10.1. In particular, the authors consider an AV moving on the road in this work. The AV is equipped with two functions, i.e. data transmission and radar sensing. At a time, it is assumed that the AV can use one of the two functions to transmit data or detect unexpected events (e.g. detect a car hidden behind a truck as shown in Figure 10.1). In this case, the MDP framework is used to model the optimal decision problem for the AV. Specifically, the system states are defined based on states of the AV and its surrounding environment, including the road state, the weather state, the AV's speed state, and the surrounding object state. Based on all the information, the AV will decide to make appropriate actions accordingly. For example, the radar function should be used when the

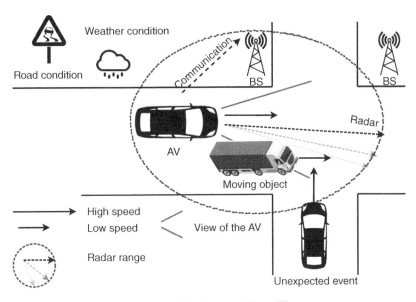

Figure 10.1 The system model of iRDC. Adapted from [3].

AV is in an environment with bad weather conditions and many obstacles around.

However, in practice, the environment information is usually not available in advance because it may vary dramatically depending on many factors. Therefore, DRL is an effective solution that allows the AV to learn the properties of the surrounding environment and quickly find its optimal policy to maximize its performance. Specifically, by making decisions (e.g. using a radar or data communication function) and observing the results received (e.g. the number of unexpected events detected or the number of packets successfully transmitted), the AV can gradually learn properties of the environment and eventually find the optimal operating policy to maximize its performance. Simulation results also show that DRL can enable the AV to find the optimal solution quickly, and thus jointly optimizing the data transmission and radar-sensing performance (i.e. enhance the user safety for the AV).

Although [3] suggested an effective solution for using DRL for JRDC systems, there are still a few limitations that can be further investigated and improved in the future works in this topic. First, we can observe that AVs often travel in large areas, and they can even frequently move into completely new environments, for example from a city to another city, from a state to another state. Therefore, every time when an AV has to move to a new environment, it needs to learn from scratch. This leads to an ineffective learning processes. Therefore, transfer learning [4] can be an effective solution to solve this problem. For example, in [5], the authors introduced a solution that allows "transfer knowledge" among AVs, so that they can reuse useful knowledge from others when they have to move to new places. We believe that this direction will be very useful in the context of AVs, and it should be further investigated for future works.

Another challenge that we can explore for future work is to consider using DRL for JRDC systems that use both functions on the same waveform. Specifically, in [3, 5], the authors considered the case when the AVs can only use one of the two functions at a time. However, many AV systems can allow the use of both functions at the same time on the same transmission waveform [6]. In this case, a few potential directions that can be further investigated. For example, DRL can be used to optimize power resources at the transmitters. Alternatively, DRL can be used to find an optimization policy for the number of preambles for transmitted waveforms [7].

10.1.2 Ambient Backscatter Communications

Ambient Backscatter (AB) is a novel wireless communication technology [8, 9]. Unlike traditional wireless communication, AB allows wireless devices to transmit data without the need of using active RF components for transmissions. The basic principle here is the use of radio signals in the surrounding environment as a medium to transmit data between wireless devices. Specifically, we consider a specific scenario that involves a wireless communication device (e.g. a mobile base station or a television base station), denoted as BS, broadcasting RF signals to the surrounding environment. In the BS communication range, we consider a transmitter and a receiver, and they are both capable of receiving the signal broadcast from the BS. In this case, when the BS transmits signals, both the transmitter and receiver can receive the signals at the same time. Thus, to transmit data to the receiver, the transmitter only needs to backscatter modulated signals to the receiver.

Let us look at an example in Figure 10.2. The incident signal is a signal broadcast by the BS. The transmitter is equipped with a backscatter circuit with impedance switching circuit. When an incident signal is received, the transmitter will control the impedance switching circuit to either backscatter the incoming signal (i.e. generate a reflect signal) or will absorb the incoming signal (i.e. all incoming signals will be absorbed, and we will have no reflect signal in this case). In this way, by controlling the impedance switching circuit, we can generate a sequence of reflect signal (modulated signal). At the receiver, the modulated signals will be used to decode information. For example, when the transmitter generates a reflect signal, it can decode as a bit "1" is transmitted. Conversely, if the transmitter does not generate a reflect signal, it can decode as a bit "0" is

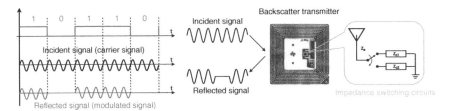

Figure 10.2 Principle of operation of ambient backscatter communication. Adapted from: https://home.cse.ust.hk/~qianzh/research/research%20-%20communication.htm

transmitted (as illustrated in Figure 10.1). As a result, by continuously controlling the operation of the impedance switching circuit, the transmitter is able to transmit a sequence of bits "0" and "1" to the receiver.

As described above, the operating principle of AB is very simple and the implementation of AB systems is also very easy since only one impedance switching circuit at the transmitter and one AB decoder at the receiver are used. Therefore, AB finds a lot of practical applications, especially IoT applications where IoT devices are often limited by communication capacity, energy usage, and limited spectrum resources. This is also the reason why AB is considered as one of 10 breakthrough technologies by the MIT Technology Review [10] and expected to bring us closer to the IoT era [11].

Despite its advantages over conventional wireless communication, there are some challenges in using AB technology in wireless communication networks. Specifically, because AB's operating principle is based on RF signals in the surrounding environment; therefore, the performance of AB communication networks depends a lot on the received RF signals. The problem here is that the signals from the BS are very dynamic and cannot be controlled by the backscatter devices. Therefore, how to control the backscatter devices in the presence of uncertainty of the wireless environment is a big challenge.

To solve the above problem, in [12], the authors introduced a DRL-based intelligent communication method to control transmitter operations. Specifically, in [12], the authors considered a scenario (similar to the one described above), i.e. consisting of one BS broadcasting RF signals to the surrounding environment and an AB communication system with one transmitter (backscatter device) and one receiver (both are located in the BS's broadcasting area). The transmitter has a data storage (represented by a data queue) and one energy storage (represented by an energy queue). The data collected from the surrounding environment will be stored in the data queue, and the harvested energy will be stored in the energy queue. The transmitter is equipped with a backscatter circuit (as described in Figure 10.3). Here, the transmitter can choose one of the following operation modes (i) if the BS is broadcasting signals, the transmitter can choose the

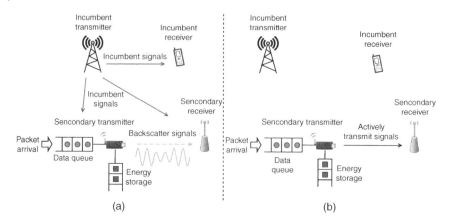

Figure 10.3 RF-powered ambient backscattering system model. (a) Incumbent channel is busy and (b) incumbent channel is idle. Adapted from [12].

backscatter mode to transmit the information to the receiver, (ii) if the BS is broadcasting signals, the transmitter can choose the energy harvesting mode, (iii) if the BS is not broadcasting signals, the BS can choose the active transmission mode, i.e. use energy in the energy queue to actively transmit data to the receiver. There is an interesting point we would like to highlight here. Specifically, the backscatter circuit of the transmitter can be used for both the backscatter mode and energy-harvesting mode because this circuit can absorb or reflect signals. Interested readers can refer to [13] for details.

Based on the model described above, the authors modeled the operational control of the transmitter using the MDP framework. Specifically, the system states include the states of the transmitter (i.e. the number of data packets in the data queue and the number of energy units in the energy queue) and the state of the medium (i.e. the BS is broadcasting signals or not). Then, RL can be used to find the optimal operation for the transmitter (e.g. backscatter data, harvest energy, or actively transmit data) with the primary goal of maximizing the long-term average throughput for the system under the dynamic and uncertainly of the surrounding wireless transmission environment as well as the incoming packet rate. The results show that the proposed framework allows the average throughput up to 50% and reduces the blocking probability and delay up to 80% compared with the conventional methods. Obviously, this work can be extended by using DRL to not only improve the learning rate for the transmitter, especially when the communication environment is constantly changing. Alternatively, considering both backscatter signals and harvesting energy from different channels at the same time could be a potential research direction in this topic.

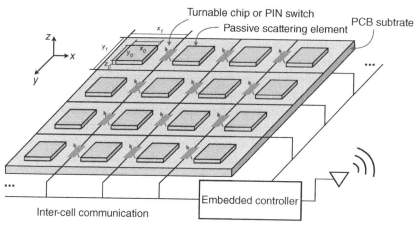

Figure 10.4 RIS architecture. Adapted from [14].

10.1.3 Reconfigurable Intelligent Surface-Aided Communications

Reconfigurable Intelligent Surface (RIS)-aided communication has recently emerged as a new communication model with many advantages and applications in future wireless networks. The basic idea of RIS communications is the use of a smart surface that is able to control incoming signals. Specifically, the smart surface consists of many reflective panels placed side-by-side on the same plane as illustrated in Figure 10.4. Each of these reflectors will be controlled by an electrical circuit and all of them allow the reflectors to modulate the reflected signals, for example amplifying them or adjusting their phases. In addition, all these reflectors will be connected and controlled by an embedded controller. This will allow the reflections on the reflectors to take place in a continuous, uniform, and highly efficient manner. In this way, since we can simultaneously control a large amount of reflected signal on this surface to the receiver, we can virtually "change" the transmission medium. In other words, we are able to change the properties of the channel to achieve the desired effect on the receiver side. For example, we can amplify the signal, and thereby increase the amount of useful signal received at the receiver. Therefore, the use of RIS communications opens up completely new applications in wireless communication technology, such as smart building, smart transport systems, and military [14].

Here, we would like to note a basic difference between AB communications [8] and RIS-based communications. The common point of these two types of communication is that they both use the same device to cause changes in the wireless transmission environment. For example, with AB communications, we use a device called a backscatter tag which is capable of reflecting the received signal to the surrounding environment. In the case of RIS-aided communications, we use a reflector to reflect the wave signals transmitted to it. Although both types of communication work in a similar way, the biggest difference here is the way of reflection. In particular, the reflective behavior of AB communications is "passive," while the reflective behavior of RIS-based communications is "active." To be more specific, with AB communications, backscatter tags have only two modes, i.e. backscatter (i.e. reflect received signal) or absorb (i.e. absorb received signal). However, in RIS-based communications systems, the reflective surfaces can not only reflect the received signal but also determine the directions and phases of the reflected signals. Therefore, reflection on RIS-based communication systems is effective in engineering the propagation environment. While AB communications are often deployed in IoT systems that require low cost and energy saving, RIS-based communication systems are often used in large systems (e.g. mobile systems) with high deployment costs to achieve high efficiency.

RIS-based communication systems face a number of challenges related to their practical implementation. In particular, controlling the activity of reflectors in RIS-based communication systems will be much more complicated than in AB communication systems since we need not only to decide whether to reflect or not but also to control the direction and phase of the reflected signal as well. In particular, after receiving the incoming waveform, the reflecting surface needs to find the direction to direct the reflected transmission beam as well as the phases for the reflecting signals to the receiver. This is a challenging problem impacted by many factors such as the properties of the propagation channel and the power and direction of the incoming waves. DRL can be used as a solution technique to tackle this problem.

In [15], the authors proposed a solution using DRL to solve the problem of optimal performance control for RIS-based communication systems. Specifically, in [15], the authors considered a system with one BS, one RIS, and a set of K users, as illustrated in Figure 10.5. It is assumed that the direct transmission from the BS to the users is blocked (e.g. due to large obstructions such as buildings). In this case, the BS can only transmit data to the users through the RIS. The task here is to control the signals received from the BS at the RIS to maximize the signal received by the users. To solve this problem, the authors first used an MDP framework to control the RIS's operation dynamically based on the actual information received. Specifically, the authors first defined the system states including the transmission power of the BS at the present time, the received power of the users at the present time, the channel state information of the environment (i.e. between the BS and the RIS and between the RIS and the users), and the action the RIS did in the previous time slot. Accordingly, the RIS can optimize its operations (i.e. optimize the transmit beamforming matrix together with the phase shift matrix of the reflectors at the same time) to maximize the

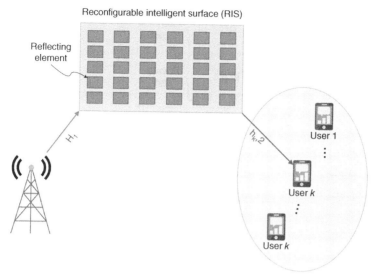

Figure 10.5 RIS system model. Adapted from [15].

long-term average reward (i.e. sum-rate capacity) of the system. The simulation results show that DRL is an effective solution to enable the system to learn the properties of the environment, thereby optimizing the system performance quickly.

In addition to the problem of phase shift control and beamforming at the RIS as mentioned above, there are other potential problems in RIS-based communication systems for which DRL can potentially provide effective solutions. For example, in a scenario where there are multiple RISs in a cell or distributed across the cells, the activities of the RISs can be controlled efficiently in a centralized way at the same time (e.g. through a server to control the RISs simultaneously by using a DRL solution). Alternatively, they can be controlled in a distributed way (i.e. each RIS can be considered as an agent and it will automatically make its own decisions based on its internal information as well as feedback information from the environment after it makes decisions).

10.1.4 Rate Splitting Communications

Multiple access is one of the fundamental problems in wireless communication networks. However, this problem becomes very difficult when the number of wireless devices becomes very large. Rate-Splitting Multiple Access (RSMA) has been introduced as a new type of communication solution that is expected to partially solve the multiple access problem in next-generation communication networks, i.e. 5G and beyond.

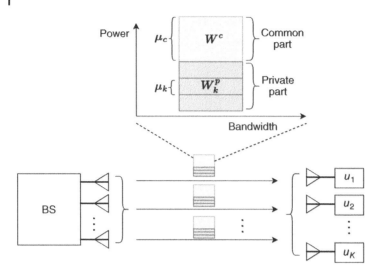

Figure 10.6 RSMA network model.

The main idea of RSMA is to split the messages and then combine the parts of the messages in an optimal way to maximize performance on the channel as well as to minimize interference on the receiver side (e.g. mobile users). Let us take a look at a specific scenario consisting of a transmitter, i.e. the BS equipped with M antennas, and K receivers, i.e. mobile users (MUs), each equipped with a single antenna, as illustrated in Figure 10.6. The BS is assumed to have K messages $W = \{W_1, \ldots, W_K\}$, and they will be delivered to the corresponding K users. However, before transmitting these messages to MUs, these messages will be split into two parts, which are private and common parts. All common parts of users will be combined together, while private parts of users will be preserved and transmitted separately. In this case, we will have a total of $K + 1$ messages, K private messages corresponding to K users and one joint big common message for all the users. In other words, we will have a total of $K + 1$ data streams, denoted by $\{s_c, s_1, \ldots, s_K\}$ corresponding to the common and $K + 1$ private messages. In this way, we can determine the signal transmitted by the BS to these K users as follows:

$$\mathbf{x} = \sqrt{\mu_c P_t}\mathbf{w}_c s_c + \sum_{k=1}^{K} \sqrt{\mu_k P_t}\mathbf{w}_k s_k, \tag{10.1}$$

where $\mathbf{w}_c \in \mathbb{C}^{M \times 1}$ and $\mathbf{w}_k \in \mathbb{C}^{M \times 1}$ are the precoding vectors of the common and private messages, respectively. μ_c and μ_k are the power allocation coefficients, i.e. the ratios between the transmission power allocated for the common and private messages to the total transmission power P_t, respectively.

Here, we would like to note that the total power allocated for all messages must be lower or equal the total power of the BS, i.e. $\mu_c + \sum_{k=1}^{K} \mu_k \leq 1$. Furthermore, if we denote $\mathbf{y}_k = \mathbf{h}_k^H \mathbf{x} + n_k$ as the received signal at user u_k, n_k as the noise at user u_k and $\mu = [\mu_c, \mu_1, \mu_2, \ldots, \mu_k, \ldots, \mu_K]$ as the power allocation coefficient vector, we can derive the SINRs of common and private messages as follows:[1]

$$\gamma_k^c(\mu) = \frac{\mu_c P_t |\mathbf{h}_k \mathbf{w}_c|^2}{\sum_{j=1}^{K} \mu_j P_t |\mathbf{h}_k \mathbf{w}_j|^2 + 1},$$

$$\gamma_k^p(\mu) = \frac{\mu_k P_t |\mathbf{h}_k \mathbf{w}_k|^2}{\sum_{j \neq k} \mu_j P_t |\mathbf{h}_k \mathbf{w}_j|^2 + 1}. \tag{10.2}$$

Given the power allocation coefficient vector μ and common rate vector $\mathbf{c} = [C_1, C_2, \ldots, C_K]$, the optimization problem is then defined as follows:

$$\max_{\mu, \mathbf{c}} \quad R_{sum}(\mu, \mathbf{c}), \tag{10.3a}$$

$$\text{s.t.} \quad \mu_c + \sum_{k=1}^{K} \mu_k \leq 1, \tag{10.3b}$$

$$\sum_{k=1}^{K} C_k \leq R_c, \tag{10.3c}$$

$$C_k + R_k \geq Q_k, k \in \mathcal{K}, \tag{10.3d}$$

$$\mathbf{c} \geq 0. \tag{10.3e}$$

In (10.3)

$$R_k(\mu) = \log_2 \left(1 + \gamma_k^p(\mu) \right), \forall k \in \mathcal{K}, \tag{10.4}$$

and

$$R_c(\mu) = \min_{k \in \mathcal{K}} \left\{ \log_2 \left(1 + \gamma_k^c(\mu) \right) \right\}. \tag{10.5}$$

Optimizing (10.3) is challenging due to the dynamic and uncertain nature of the channel since the channel gain $\mathbf{h}_k$ between the BS and user u_k varies over time, and the channel state distribution is unknown by the BS. Unlike conventional multiple access schemes, splitting messages into different parts makes the problem even more challenging because the power needs to be allocated in a way that all the messages are decodable. DRL can be an effective tool to address this problem.

In [16], the authors introduced an MDP framework to formulate the dynamic resource allocation for the BS and proposed an effective DRL solution using

---

1 Here, it is assumed that the noise power is normalized for the sake of simplicity.

proximal policy optimization algorithm to find the optimal policy for the BS. First, the authors define the state space of the system as the feedback signals/messages received from all the users regarding their received common and private parts. These feedback messages are important information about the channel state of the system which may not be observable by the BS due to dynamic and uncertainty of wireless channels. Therefore, based on this information, the BS can make optimal decisions, i.e. optimal power allocation for the transmitted common and private messages, with the aim to optimize the total sum-rate of transmitted messages (while ensuring the QoS is imposed in constrains in (10.3) for the users). To do so, the authors designed a utility function that consists of two components, the first one is the reward for the total sum-rate it receives, and the second one is the penalty if the QoS for the users is violated. In addition, since actions are continuous variables, the authors developed an intelligent algorithm based on the proximal policy optimization algorithm [17], which allows the BS to gradually find the optimal power allocation policy for the messages sent to the users. The results show that the proposed learning algorithm can help the BS to quickly find the optimal policy and it can outperform the baseline methods.

10.2 Advanced DRL Models

10.2.1 Deep Reinforcement Transfer Learning

DRL is a rather special learning algorithm because unlike traditional ML algorithms where data are available for the learning process, DRL requires a learner to collect data through interacting with the surrounding environment, before it can analyze and extract information from the collected data. Therefore, the learning process of DRL is often very slow compared to the learning process of traditional ML algorithms, e.g. supervised learning [18]. As a result, Transfer Learning (TL) techniques have recently been introduced as an efficient way that can significantly speed up the learning processes for DRL.

At a high level, TL is a solution approach that allows knowledge extraction from learning tasks from one source domain to improve learning for one learning task in the target domain, as illustrated in Figure 10.7. Here, the main idea of transfer knowledge comes from the idea that the previously learned knowledge may help for the knowledge that we need to learn later. For example, a person who has learned the violin before may learn guitar much faster than another person who has never used musical instruments before [4, 19, 20]. TL can be used for DRL by transferring knowledge learned from previous agents to new learning agents in the future. For example, the mining tasks of robots previously performed on earth can be aggregated and transferred to robots with the task of searching for minerals on other planets, e.g. the moon and mars, in the future.

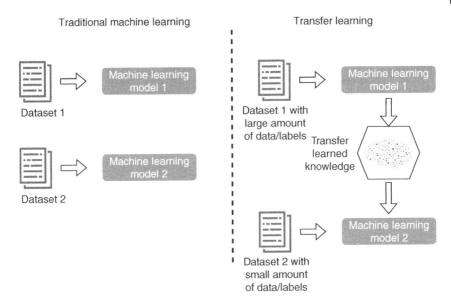

Figure 10.7 Transfer learning model.

In general, there are three main techniques in TL, which are inductive TL, transductive TL, and unsupervised TL. Each technique has different properties and therefore will be used for different purposes as summarized in Table 10.1 [4]. For example, if both source and target domains have labeled data, inductive TL can be very useful for multitask learning (e.g. a robot wants to learn multiple tasks, such as transmitting data, exploring the environment, and searching natural resources). Or if the target domain has no labeled data, but the robot at the target domain has the same task (e.g. transmitting data, exploring the environment and searching natural resources as in the example above), transductive TL can be used.

Theoretically, all of the above techniques can be used in the context of DRL. However, there are some important notes for using TL techniques in DRL scenarios. In particular, one of the most distinctive features of DRL, which differs from other conventional ML algorithms, is that learning problems are often modeled using the MDP framework. In an MDP, an agent takes actions and observes the results from the environment. The agent's goal is to find an optimal policy that optimizes its objective function in a long term. Therefore, there are four main strategies for TL in DRL as follows.

10.2.1.1 Reward Shaping

Reward shaping is one of the effective solutions to help the target MDP agent quickly find the optimal policy by adjusting its reward function. Specifically, the

Table 10.1 Transfer learning techniques.

TL strategy	Related research areas	Source and target domain	Source domain labels	Target domain labels	Source and target tasks
Inductive TL	Multitask Learning	The same	Available	Available	Different
			Available	Available	But related
Inductive TL	Self-taught Learning	The same	Unavailable	Available	Different
			Unavailable	Available	But related
Transductive TL	Domain Adaptation, Sample Selection Bias and Covariate Shift	Different But related	Available	Unavailable	The same
Unsupervised TL		Different but related	Unavailable	Unavailable	Different But related

main idea here is to extract the knowledge learned from the sources of MDP agents to influence the reward function of the target MDP agent, thereby helping the target agent quickly find the best actions to optimize its policy. For example, if at a source MDP, we know that the optimal action at state s is a_1, then in this case, when the target MDP agent is also at the same state and in a similar learning environment as that of the source MDP agent, we can adjust the reward function of the target MDP to increase the reward if it selects action a_1. In this way, we can allow the target MDP agent to significantly speed up the trial-and-error period in order to converge quickly to the optimal policy. However, a limitation of this method is that choosing an enhanced reward function for the target MDP agent is very difficult because it depends a lot on environmental factors of both source and traction agents.

10.2.1.2 Intertask Mapping

This is an extended version of reward shaping strategy which is used to help overcome the limitations of the reward shaping strategy. Specifically, in reward shaping, it often requires the states and actions of the MDP source and the target MDP to be very similar, and thus we can adjust the reward function of the target MDP based on the experiences drawn from the MDP source. However, in reality, there will be many cases where the source MDP and target MDP are not the same in both states and actions. Furthermore, the tasks of sources and target MDPs may not be the same. For example, the source MDP agent wants to minimize its transmission delay, while the target MDP agent wants to minimize its energy consumption. Therefore, the intertask mapping strategy can be used to find the relationship between the source and target MDP, thereby mapping the learned knowledge from the MDP source to the MDP target. Some common mapping methods include, for example, only state space (all or part of state space) [21–23], or representation of dynamic transition [24].

10.2.1.3 Learning from Demonstrations

A demonstration is defined as an experience of an agent's interaction with its surroundings. Specifically, we can define a demonstration by a set of four components including current state, action, next state, and reward, which is usually denoted by (s, a, r, s'). Demonstrations (or experiences) from source domains (source MDPs) can then be passed to a target domain (target MDP) to improve agent learning of the target MDP. This transfer experiences can take place online (i.e. the target MDP will receive experiences from other source MDPs and then mix them with the experiences it obtains in the experience pool to learn its optimal policy in a real-time manner) or offline (i.e. the target MDP will use transferred experiences to train its neural network before using the trained model for learning process). Here, it is important to note that the experiences when passing from source MDPs to the target MDPs can be nearly optimal, suboptimal, optimal, or even nonoptimal. Therefore, it is very important to choose experiences to transfer to the target MDP agent. In addition, another point we need to note here is that experiences from the source MDPs must have the same properties as that of the target MDP. For example, the MDP sources and the target MDP must have the same state and action spaces, otherwise, the target MDP agent cannot learn useful knowledge from the source MDP agents' experiences. More importantly, if the source MDP agents' experiences are too different from those of the target MDP agent, the target MDP agent even cannot find its optimal policy.

10.2.1.4 Policy Transfer

This strategy is one of the most widely used strategy in deep reinforcement transfer learning. The idea here is quite simple. An optimal policy obtained by a source MDP agent will be transferred to a target MDP agent for further processing. In the context of DRL, an optimal policy is a deep neural network after being fully trained at the source MDP. Therefore, transferring policies means transferring trained deep neural networks from the source MDP. There are two commonly used methods, i.e. Policy Distillation and Policy Reuse [4].

- *Policy distillation*: In this method, the optimal policies from the source MDP agents are distilled to form a new policy for the target MDP to use. Since in DRL, policy transfer is essentially the transfer of trained models from the source MDPs, distilling policies here is distilling information from deep neural networks. However, there is one point that we need to note here. In the process of distilling policies from the source MDPs for the target MDP, we need to carefully control the weights of transferred trained deep learning models from the source MDPs. Different source MDPs may have different learning environments, and thus distributions of their trained models may also be different. Thus, controlling transferred weights for different trained models from the source MDPs is still remaining the big question.

- *Policy reuse*: Unlike policy distillation strategy which seeks to distill knowledge from trained models and from the source MDPs, in policy reuse strategy, a target MDP will reuse optimal policies from source MDPs directly. In particular, we will use a probability function to determine the policy of the target MDPs as follows:

$$Prb(\pi_i) = \frac{e^{tG_i}}{\sum_{j=0}^{N}(e^{tG_j})}, \tag{10.6}$$

where N is the number of policies and t is the learning time (that will increase over time). Here, it is noted that G_j expresses the performance of the MDP agent j, and G_0 is the current policy of the target MDP agent. To evaluate performance of a policy, Q-function can be used [25, 26].

10.2.1.5 Reusing Representations

Another way to reuse trained learning models from source MDPs for a target MDP is through the representation strategy [27–29]. Specifically, this strategy allows the target MDP to reuse trained deep neural networks from the source MDPs as input layers in its new deep neural network. Here, different from the policy transfer strategy where we directly use the trained deep neural networks from the source MDPs, in the reusing representation strategy, the trained deep neural networks from the source MDPs are used as feature extractors to produce new representations for the new deep neural network of the target MDP. This strategy allows solving the problem by finding the right mapping function or the right weights to combine trained models like the solutions above. However, it has a problem that integrating neural networks from the source MDPs to the target MDP will lead to a very big deep neural network and thus result in a very long and energy-consuming process. This thus limits applications of reusing representation strategy in practice, especially in the context of wireless networks.

In Table 10.2, we summarize all potential strategies for using TL in DRL. In this table, $\mathcal{M}_S$ and $\mathcal{M}_T$ denote the source and target MDP, respectively. $\mathcal{S}_S$, $\mathcal{A}_S$, and $\mathcal{R}_S$ are state space, action space, and reward function of source MDP $\mathcal{M}_S$. Similarly, $\mathcal{S}_T$, $\mathcal{A}_T$, and $\mathcal{R}_T$ are state space, action space, and reward function of target MDP $\mathcal{M}_T$.

Table 10.2 Deep reinforcement transfer learning strategies.

TL-DRL strategy	MDP difference	Knowledge to transfer
Reward shaping	$\mathcal{M}_S = \mathcal{M}_T$	Reward function
Learning from experiences	$\mathcal{M}_S = \mathcal{M}_T$	Experiences, i.e. (s, a, s', r)
Policy transfer	$\mathcal{A}_S \neq \mathcal{A}_T$; $\mathcal{R}_S \neq \mathcal{R}_T$	Source agent's policy
Inter-task mapping	$S_S \neq S_T$; $S_S \neq S_T$ and $\mathcal{A}_S \neq \mathcal{A}_T$; $\mathcal{R}_S \neq \mathcal{R}_T$ and $\mathcal{A}_S \neq \mathcal{A}_T$; $S_S \neq S_T$ and $\mathcal{R}_S \neq \mathcal{R}_T$; $S_S \times \mathcal{A}_S \neq S_T \times \mathcal{A}_T$	Source MDP
Reusing representations	$S_S \neq S_T$; $S_S \neq S_T$ and $\mathcal{A}_S \neq \mathcal{A}_T$; $\mathcal{R}_S \neq \mathcal{R}_T$ and $\mathcal{A}_S \neq \mathcal{A}_T$; $S_S \neq S_T$ and $\mathcal{R}_S \neq \mathcal{R}_T$; $S_S \times \mathcal{A}_S \neq S_T \times \mathcal{A}_T$	Representations of states or transition dynamics

10.2.2 Generative Adversarial Network (GAN) for DRL

In the learning process of an RL agent, there are two main processes that are exploration and exploitation. On the one hand, exploration is the process by which the agent attempts to explore the world around him by selecting different actions to get multiple views and results of his actions. On the other hand, exploitation is the process of focusing on choosing the actions that bring the most benefit to the agent (after exploring other actions for a sufficient time). Here, we can see that the exploration process plays a very important role in the agent's learning process. The reason is, if the exploration is too short, it will not be able to give the agent a completed view of the surrounding that it needs to learn. However, if the exploration process is too long, it will shorten the exploitation time, and thus the benefit for the agent will be reduced. In the reality of wireless networks, exploration can have significant effects on the quality of service. This is because the exploration data is not enough to train the deep learning model effectively. Therefore, the authors in [30] developed a Generative Adversarial Network (GAN) model that can help to generate a lot of artificial data to speed up the agent's initial learning process in a better way.

Specifically, the authors proposed an enhanced GAN structure integrated into an RL agent to generate additional artificial data for the RL agent, as illustrated in Figure 10.8. In the process of interacting with the surrounding environment, the agent will acquire experiences, e.g. represented as a set of (current state, action, next state, and reward). This experience is also known as a data sample. This data will then be fed into the proposed GAN model to generate a new data sample (called an artificial data sample). All samples (both actual and artificial) will be

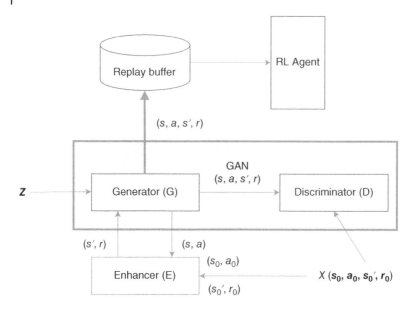

Figure 10.8 Enhanced GAN structure for DRL.

put into a replay buffer for storage. All this data will then be used to train the deep learning model for the reinforcement learning agent. In this way, we can generate a lot of artificial data for the reinforcement learning agent to learn. Although artificial data cannot be as good as the actual data, they have similar properties to the actual data. Thus, the learning process will be significantly improved, especially for deep neural networks which only can obtain good performance when they have sufficient data to train.

10.2.3 Meta Reinforcement Learning

Reinforcement learning is known as an effective solution in learning through interacting with the surrounding environment. However, a disadvantage of this approach is that after the learning process, the agent can only learn a specific policy that is appropriate for that environment. If that policy is used in a new environment (even if the new environment is very similar to the old one learned), it also may not be able to achieve the desired performance. As a result, meta reinforcement learning [31, 32] is emerging as an effective solution that is expected to solve this problem.

Theoretically, meta-learning can be understood as a process of "learn how to learn." More specifically, in the context of reinforcement learning, the use of meta-learning helps an agent learn how it can obtain that optimal policy, rather than just learning to find its optimal policy, like the traditional way of reinforcement learning. Therefore, basically, the learning process of a meta reinforcement learning would be exactly the same as the learning process of a reinforcement learning, i.e. an agent would observe the current state, make an action, and then obtain the next state together with an associated reward. However, the only one difference here is the way the agent updates the optimal policy for the agent. Specifically, in reinforcement learning, the optimal policy π_{s_t} will be represented through an allocation function over action space $\mathcal{A}$. However, in meta-reinforcement learning, an agent's policy will cover both the last reward and the last action taken, i.e. $\pi_\theta(a_{t-1}, r_{t-1}, s_t) \to \mathcal{A}$. In this way, the agent can "internalize the dynamics between states, rewards, and actions in the current MDP and adjust its strategy accordingly" [31]. Therefore, if an agent needs to learn in a similar environment, the agent's policy can quickly adapt and can be effectively reused for learning. In the context of using Meta-learning for RL tasks, model-agnostic meta-learning algorithm (MAML) [33] and Reptile [34] are among the most effective methods that show good generalization performance on new tasks for RL agents. More potential research can be considered the use of meta-learning on deep neural networks for RL agents.

10.3 Chapter Summary

In this chapter, we have discussed some potential research directions. In particular, we have discussed applications of DRL algorithms to address emerging problems in wireless networks (e.g. joint radar and data, ambient backscatter communications, intelligent reflecting surface, and rate splitting communications) and potential advanced DRL techniques (e.g. deep reinforcement transfer learning, GAN deep reinforcement learning, and meta-reinforcement learning). For each topic, we have briefly explained the idea and provided a case study to discuss its potential. Although studies on these topics are in their infancy, we believe they are very promising research directions that can not only address some of the issues discussed in Chapter 9 but also provide potential research directions for future wireless networks.

References

1 N. C. Luong, X. Lu, D. T. Hoang, D. Niyato, and D. I. Kim, "Radio resource management in joint radar and communication: A comprehensive survey," *IEEE Communications Surveys & Tutorials*, vol. 23, pp. 780–814, 2021.
2 F. Liu, C. Masouros, A. P. Petropulu, H. Griffiths, and L. Hanzo, "Joint radar and communication design: Applications, state-of-the-art, and the road ahead," *IEEE Transactions on Communications*, vol. 23, pp. 3834–3862, 2020.

3 N. Q. Hieu, D. T. Hoang, N. C. Luong, and D. Niyato, "iRDRC: An intelligent real-time dual-functional radar-communication system for automotive vehicles," *IEEE Wireless Communications Letters*, vol. 9, pp. 2140–2143, 2020.

4 C. T. Nguyen, N. V. Huynh, N. H. Chu, Y. M. Saputra, D. T. Hoang, D. N. Nguyen, Q.-V. Pham, D. Niyato, E. Dutkiewicz, and W.-J. Hwang, "Transfer learning for future wireless networks: A comprehensive survey," *arXiv preprint arXiv:2102.07572*, 2021.

5 N. Q. Hieu, D. T. Hoang, D. Niyato, P. Wang, D. I. Kim, and C. Yuen, "Transferable deep reinforcement learning framework for autonomous vehicles with joint radar-data communications," *arXiv preprint arXiv:2105.13670*, 2021.

6 D. Ma, N. Shlezinger, T. Huang, Y. Liu, and Y. C. Eldar, "Joint radar-communication strategies for autonomous vehicles: Combining two key automotive technologies," *IEEE Signal Processing Magazine*, vol. 37, pp. 85–97, 2020.

7 N. H. Chu, D. N. Nguyen, D. T. Hoang, Q.-V. Pham, K. T. Phan, W.-J. Hwang, and E. Dutkiewicz, "AI-enabled mm-Waveform configuration for autonomous vehicles with integrated communication and sensing," *IEEE Internet of Things Journal*, 2023.

8 D. T. Hoang, D. Niyato, D. I. Kim, N. V. Huynh, and S. Gong, *Ambient backscatter communication networks*. Cambridge University Press, 2020.

9 N. V. Huynh, D. T. Hoang, X. Lu, D. Niyato, P. Wang, and D. I. Kim, "Ambient backscatter communications: A contemporary survey," *IEEE Communications Surveys & Tutorials*, vol. 20, no. 4, pp. 2889–2922, 2018.

10 M. T. Review, "10 breakthrough technologies 2016," 2017.

11 C. Dunne, "Ambient backscatter brings us closer to an Internet of Things," 2017.

12 N. V. Huynh, D. T. Hoang, D. N. Nguyen, E. Dutkiewicz, D. Niyato, and P. Wang, "Optimal and low-complexity dynamic spectrum access for RF-powered ambient backscatter system with online reinforcement learning," *IEEE Transactions on Communications*, vol. 67, no. 8, pp. 5736–5752, 2019.

13 D. T. Hoang, D. Niyato, P. Wang, D. I. Kim, and Z. Han, "Ambient backscatter: A new approach to improve network performance for RF-powered cognitive radio networks," *IEEE Transactions on Communications*, vol. 65, no. 9, pp. 3659–3674, 2017.

14 S. Gong, X. Lu, D. T. Hoang, D. Niyato, L. Shu, D. I. Kim, and Y.-C. Liang, "Toward smart wireless communications via intelligent reflecting surfaces: A contemporary survey," *IEEE Communications Surveys & Tutorials*, vol. 22, no. 4, pp. 2283–2314, 2020.

15 C. Huang, R. Mo, and C. Yuen, "Reconfigurable intelligent surface assisted multiuser MISO systems exploiting deep reinforcement learning," *IEEE Journal on Selected Areas in Communications*, vol. 38, no. 8, pp. 1839–1850, 2020.

16 N. Q. Hieu, D. T. Hoang, D. Niyato, and D. I. Kim, "Optimal power allocation for rate splitting communications with deep reinforcement learning," *IEEE Wireless Communications Letters*, vol. 10, no. 12, pp. 2820–2823, 2021.

17 J. Schulman, F. Wolski, P. Dhariwal, A. Radford, and O. Klimov, "Proximal policy optimization algorithms," *arXiv preprint arXiv:1707.06347*, 2017.

18 Y. C. Eldar, A. Goldsmith, D. Gündüz, and H. V. Poor (eds.), *Machine learning and wireless communications*, Cambridge University Press, Aug. 2022.

19 S. J. Pan and Q. Yang, "A survey on transfer learning," *IEEE Transactions on Knowledge and Data Engineering*, vol. 22, no. 10, pp. 1345–1359, 2009.

20 F. Zhuang, Z. Qi, K. Duan, D. Xi, Y. Zhu, H. Zhu, H. Xiong, and Q. He, "A comprehensive survey on transfer learning," *Proceedings of the IEEE*, vol. 109, no. 1, pp. 43–76, 2020.

21 A. Gupta, C. Devin, Y. Liu, P. Abbeel, and S. Levine, "Learning invariant feature spaces to transfer skills with reinforcement learning," *arXiv preprint arXiv:1703.02949*, 2017.

22 G. Konidaris and A. Barto, "Autonomous shaping: Knowledge transfer in reinforcement learning," in *Proceedings of the 23rd International Conference on Machine Learning*, pp. 489–496, 2006.

23 H. B. Ammar and M. E. Taylor, "Reinforcement learning transfer via common subspaces," in *Adaptive and Learning Agents. International Workshop on Adaptive and Learning Agents, Lecture Notes in Computer Science* (P. Vrancx, M. Knudson, and M. Grześ, eds.), vol. 7113, pp. 21–36, Springer, 2011.

24 H. B. Ammar, K. Tuyls, M. E. Taylor, K. Driessens, and G. Weiss, "Reinforcement learning transfer via sparse coding," in *Proceedings of the 11th International Conference on Autonomous Agents and Multiagent Systems*, vol. 1, pp. 383–390, International Foundation for Autonomous Agents and Multiagent Systems ..., 2012.

25 F. Fernández and M. Veloso, "Probabilistic policy reuse in a reinforcement learning agent," in *Proceedings of the 5th International Joint Conference on Autonomous Agents and Multiagent Systems*, pp. 720–727, 2006.

26 A. Barreto, W. Dabney, R. Munos, J. J. Hunt, T. Schaul, H. Van Hasselt, and D. Silver, "Successor features for transfer in reinforcement learning," *arXiv preprint arXiv:1606.05312*, 2016.

27 A. A. Rusu, N. C. Rabinowitz, G. Desjardins, H. Soyer, J. Kirkpatrick, K. Kavukcuoglu, R. Pascanu, and R. Hadsell, "Progressive neural networks," *arXiv preprint arXiv:1606.04671*, 2016.

28 C. Fernando, D. Banarse, C. Blundell, Y. Zwols, D. Ha, A. A. Rusu, A. Pritzel, and D. Wierstra, "PathNet: Evolution channels gradient descent in super neural networks," *arXiv preprint arXiv:1701.08734*, 2017.

29 C. Devin, A. Gupta, T. Darrell, P. Abbeel, and S. Levine, "Learning modular neural network policies for multi-task and multi-robot transfer," in *2017 IEEE International Conference on Robotics and Automation (ICRA)*, pp. 2169–2176, IEEE, 2017.

30 V. Huang, T. Ley, M. Vlachou-Konchylaki, and W. Hu, "Enhanced experience replay generation for efficient reinforcement learning," *arXiv preprint arXiv:1705.08245*, 2017.

31 L. Weng, "Meta reinforcement learning," 2019.

32 R. Sahay, "Learning to learn more: Meta reinforcement learning," 2020.

33 C. Finn, P. Abbeel, and S. Levine, "Model-agnostic meta-learning for fast adaptation of deep networks," in *International Conference on Machine Learning*, pp. 1126–1135, PMLR, 2017.

34 A. Nichol, J. Achiam, and J. Schulman, "On first-order meta-learning algorithms," *arXiv preprint arXiv:1803.02999*, 2018.

Index

Deep Reinforcement Learning for Wireless Communications and Networking:
Theory, Applications, and Implementation, First Edition.
Dinh Thai Hoang, Nguyen Van Huynh, Diep N. Nguyen, Ekram Hossain, and Dusit Niyato.
© 2023 The Institute of Electrical and Electronics Engineers, Inc. Published 2023 by John Wiley & Sons, Inc.

Printed and bound by CPI Group (UK) Ltd, Croydon, CR0 4YY

27/10/2024

14580669-0002